Globalization and a High-Tech Economy

Globalization and a High-Tech Economy:
California, the United States and Beyond

by

Ashok Deo Bardhan
Dwight M. Jaffee
Cynthia A. Kroll
Haas School of Business, University of California

Kluwer Academic Publishers
Boston, New York, Dordrecht

Distributors for North, Central and South America:
Kluwer Academic Publishers
101 Philip Drive
Assinippi Park
Norwell, Massachusetts 02061 USA
Telephone (781) 871-6600
Fax (781) 681-9045
E-Mail <kluwer@wkap.com>
Distributors for all other countries:
Kluwer Academic Publishers Group
Post Office Box 17
3300 AH Dordrecht, THE NETHERLANDS
Tel: +31 (0) 78 657 60 00
Fax: +31 (0) 78 657 64 74

E-Mail <services@wkap.nl>

 Electronic Services <http://www.wkap.nl>

Globalization and a High-Tech Economy:
 California, the United States and Beyond
   Bardhan, A., Jaffee, D., Kroll, C.
     p.cm.
Includes index.
ISBN 0-7923-7317-0

Dedication

To our spouses,

Arden Hall, Lynne LaMarca Heinrich, and Raka Ray

with appreciation for help and patience.

Contents

Figures and Tables

Acknowledgements

This work is an extension of an earlier research project, *Foreign Trade and California's Economic Growth*, funded by the California Policy Seminar and the Fisher Center for Real Estate and Urban Economics. Two former colleagues at the Fisher Center, David Howe and Josh Kirschenbaum, were coauthors on this earlier research and contributed much to the foundation from which this work has grown. Graduate and undergraduate student researchers provided excellent research assistance throughout the years of this research, including Ning Chen, Jonathan Ho, Jesse Kerns, Nancy Kim, Jeanine Kranitz, Ran Li, Helen Shvets, Jacqueline Tse, Jee Woo and Nan Zhou.

The authors would like to thank David Barton, Holly Brown-Williams, Kim Corley, Tom Davidoff, Lloyd Day, Rafiq Dossani, Robert Edelstein, Janie Fong, Robert Feenstra, Jeffrey Frankel, Gordon Hanson, Qaizar Hussain, Andres Jimenez, Ed Kawahara, Vinod Khosla, Arcady Khotin, Gus Koehler, Charles Leung, Jit Malik, Gina Mandy, Rosa Moller, Balaji Parthasarthy, Marjorie Pavliscak, Gary Peete, Annalee Saxenian, Branko Urosevic, Vince Valvano, Nancy Wallace, and seminar participants at UC Berkeley, Federal Reserve Bank San Francisco, Stanford University, the California Technology, Trade and Commerce Agency, and the Public Policy Institute of California for valuable suggestions. We would also like to thank the staff at the Fisher Center for Real Estate & Urban Economics, Linda Algazzali, Lorraine Girlington, Lynn Lobner, Jo Magaraci, Thomas Randle, Seanna Tise and Zee Zeleski, all of whom helped us with various aspects of this difficult task. All errors are our very own.

Preface

This work began in 1995 as a research effort to understand the role foreign trade had played and would continue to play in the recovery of the California economy from the 1991 through 1993 recession. In 1996 and 1997, the research continued, under a grant from the California Policy Seminar (since renamed the California Policy Research Center) of the University of California. The work resulted in a policy brief, a report issued by the California Policy Seminar, and seven working papers released by the Fisher Center for Real Estate and Urban Economics at the Haas School of Business, University of California, Berkeley.

Since the initial series of reports was completed, the forces of globalization have continued to transform high-tech industries, especially the "computer cluster," in California, the United States, and beyond. The computer cluster includes semiconductors, computer components and machines, and various aspects of software including systems design, data processing, and software. While some high-tech sectors, such as semiconductor production, have long taken advantage of global resources in their production processes and marketing, presently even sectors that once relied largely on domestic markets and domestic production sites have begun to move outward for both inputs to production and to find new markets.

This book is a joint undertaking of the three authors and we all share equally in any praise or blame. However, books are not written well by committee, and each of us has taken on special responsibilities as follows (alphabetical order has been used in all lists of author names in this book):

Chapter 1	Ashok Deo Bardhan, Dwight Jaffee, and Cynthia Kroll
Chapter 2	Dwight Jaffee
Chapter 3	Cynthia Kroll
Chapter 4	Ashok Deo Bardhan
Chapter 5	Ashok Deo Bardhan and Dwight Jaffee
Chapter 6	Ashok Deo Bardhan
Chapter 7	Cynthia Kroll
Chapter 8	Ashok Deo Bardhan, Dwight Jaffee, and Cynthia Kroll

Chapter 1

Introduction

In the past five decades, high-technology industry has progressed from a small part of the United States (US) manufacturing base to a major driver of both the manufacturing and services sectors of the economy. The growth and transformation of the industry has been integral to the expansion of metropolitan areas in the southern and western US and to the recovery of a number of northeastern and midwestern cities previously dependent on heavy manufacturing. As these regions became tied to high-tech sectors, their economies have also been necessarily shaped by the process of globalization, with sales, purchases, and production taking place beyond their traditional borders.

1.1 SUBJECT OF THE BOOK

This book focuses on the interaction between a high-tech economy and the forces of globalization, and the impact this interaction has on the regions in which the activities of the high-tech firms are concentrated. In this introductory chapter, we define the terms *high-tech industries*, *globalization*, and *regional concentration*, introduce the framework of our analysis, and summarize the issues raised.

1.1.1 High-Tech Industries

There is no single accepted definition of a "high-technology" industry. For our purposes, high-tech industries have four key features:

1) Innovation underlies product development in these industries; research and development activity is crucial.
2) The industries rely on large amounts of highly skilled labor or human capital, relative to total employment. Skilled labor is the key input in both the *research activity* that creates knowledge and the *design activities* that create the specific products, both hardware and software.
3) The exchange of ideas fuels the innovation and design processes, causing these activities to be concentrated in a few locations, not broadly and evenly distributed across the whole economy or country.
4) The routine production processes in these industries do not require intense intellectual networking and interaction, allowing the production to be geographically separated from the design and management functions. Thus, high-tech industries are well-suited to a production process that occurs in multiple locations, indeed multinationally.

Definitions of high-tech industrial sectors generally include the first two points in the above list. For example, almost twenty years ago, *Monthly Labor Review* compared definitions based on the proportion of scientific and technical workers, the relative level of research and development investments, and a combination of the two (Riche, Hecker and Burgan, 1983). These and other authors also refer to "product sophistication" as a characteristic distinguishing high-tech products from other manufactured goods or other types of services (Riche, Hecker and Burgan, 1983; Vinson and Harrington, 1979; and Markusen, Hall and Glasmeier, 1986). Measurement of product sophistication, however, is highly subjective, and many studies rely on worker classifications and research expenditures to distinguish high tech from other products.

The *Monthly Labor Review* article shows how widely the results may vary for different definitions. Only six sectors defined by 3-digit Standard Industrial Classification (SIC) codes (drugs, computers, communications equipment, electronic components, aircraft and parts, and guided missiles and space vehicles) are "high-tech" under a definition of "at least twice the average ratio of R&D expenditures to total sales", while 48 sectors, including heavy construction and hydraulic cement, are included in a definition based on the proportion of "scientific and technical workers".

Among the many alternatives, we focus in this book on a definition relevant to industries clustered around computer manufacturing and services. This approach goes beyond just manufacturing sectors (in contrast to the R&D definition), but narrows the industry focus relative to a definition based on the proportion of scientific and technical employees. In Chapter 2, we designate the precise industry codes that we adopt in our statistical analysis of the high-tech sectors.

1.1.2   Globalization and Trade

Globalization is also a term that has many definitions. For example, Mittleman (2000), distinguishes two main categories of definitions:

> The first of these is to point to an increase in interconnections, or interdependence, a rise in transnational flows, and an intensification of processes such that the world is, in some respects, becoming a single place,

and

> A second cut is more theoretical and emphasizes the compression of time and space. [...] Space is increasingly dislocated from place, and networked to other social contexts across the globe. [1]

Kemp and Shimomura (1999), distinguish between statistical and conceptual definitions:

> The economy is internationalized in the statistical sense if the ratio of world trade to world output increases or if the ratio of foreign investment to world output increases, or if the information available to households and firms becomes more uniform over countries. It is internationalized in the conceptual sense if there is a loosening of restrictions which govern international trade and investment, or the international dispersion of information...[2]

In our work, we are most interested in the aspects of globalization that affect trade patterns and production networks. Thus, by globalization, we mean *the institutional developments that allow international transfers of information, people, goods, and services to be carried out rapidly and at low cost.* In a fundamental sense, globalization reverses the process through which countries previously had inhibited international trade by erecting explicit trade barriers or by creating other impediments to curtail the international flow of goods and services. Globalization thus reduces the transactions cost of international trading, including the costs both of transporting physical goods and of transferring knowledge.

---

[1] Mittelman (2000), pages 5-6.
[2] Kemp and Shimomura (1999), page 1.

1.1.3 Globalization and the High-Tech Economy

Globalization interacts with a high-tech economy in a number of unique and direct ways:

1) Globalization makes it feasible for the high-tech industry to separate the headquarter activities of research, design, and management from the actual production of the goods. In particular, it is feasible to use low-cost, foreign locations to manufacture the goods, thus further accentuating the process of international division and specialization of labor.

2) The production process can be further disaggregated, so that either hardware or software inputs may be "outsourced" to other locations, including other countries. Examples include memory production in Taiwan and Singapore and software production in India and Russia. Such outsourcing may provide direct access to final sales in foreign markets, and/or may provide low-cost intermediate inputs for further production and assembly in the United States.

3) Scarce human capital resources, such as engineering talent, can be attracted from abroad to work at the headquarters in research and design activities.

4) Immigrant networks centered in high-tech regions can further influence global trade patterns, including both access to foreign markets and foreign production in the immigrants' home countries. See Saxenian, Motoyama and Quan (2002).

5) High-tech innovations such as communications advances have made previously non-tradable sectors tradable, with worldwide production not just in basic manufacturing, but also in specialized manufacturing, software, and other high-tech services.

In summary, high-tech industries both benefit from and facilitate globalization.

1.1.4 Regional Concentration

Regional concentration or agglomeration refers to the tendency for industries to congregate geographically in a specific region, instead of spreading broadly and evenly across an entire country. Although regional concentration may exist for any industry (auto production in the Detroit region, for example), it is particularly common in high-tech industries. One result is that the economic effects of the high-tech economy tend to be aggregated in specific regions (although, of course, very important effects are also likely to be felt in the economy as a whole). Markusen, Hall and Glasmeier (1986) point out that aspects of this concentration continue even as the industry matures. They describe "a whole continuum of spatial organizational relationships" that includes:

> primary centers of basic research and development populated by entrepreneurial ventures and branches of large corporations, secondary concentrations of technical branch establishments undertaking product-line R&D as well as assembly and production, and tertiary clusters of standardized branch production and assembly facilities.[3]

They also identify five major cores of high-tech agglomeration, based on a broad definition of high tech (similar to the broader *Monthly Labor Review* definition), centered around California, Texas, Massachusetts, Illinois and New Jersey.

More recent research finds agglomeration continues to exist in high-tech sectors, often accompanied by specialization. DeVol (1999) uses measures of output and aggregated industrial activity to identify US metropolitan areas where high tech is concentrated.[4] Many of these metropolitan areas are in the five states originally identified by Markusen, Hall and Glasmeier as major cores of high-tech agglomeration. However, DeVol notes that many of the metropolitan areas most concentrated in high tech twenty years ago have become less concentrated relative to other places (as might be expected in a maturing industry). He also notes that agglomeration and dispersion patterns vary between the manufacturing and service high-tech industries.

Work by Cortwright and Mayer (2001) confirms the differential patterns among metropolitan areas with high-tech concentrations. They find that "in most high-tech regions, high-tech employment is concentrated in only a few industry segments." Innovation is concentrated within each region among a few firms in the dominant sector, and venture capital also flows disproportionately to firms in the industry of concentration.[5] Of all of the metropolitan areas they studied, San Jose (which is part of Silicon Valley) stood out for its size and its much broader representation among different high-tech sectors. The continuing dominance of the San Jose metropolitan area in high tech, despite a long period of maturing and dispersal of the industry, places California in a unique position from which to view the effects of the industry's globalization.

---

[3] Markusen, Hall and Glasmeier (1986), pages 76 and 77.
[4] DeVol (1999), Section 3.
[5] Cortright and Mayer (2001), page 1.

1.1.5   California's Computer Industry as a Key Example

California has benefited greatly from the high-tech explosion, beginning with the aerospace industry in southern and northern California, to computer products in Silicon Valley, biotechnology in San Diego, and multimedia in San Francisco. The growth has been dynamic, bringing changes to the places where it occurs and to the companies involved in the industry. In the late 20th and early 21st century, it has been characterized by two countervailing forces. The benefits of agglomeration for innovation and new product development *pull* firms into California, at the same time that the forces of globalization *disperse* production activities and firms around the world.

This book uses the computer cluster (including computer manufacturing, semiconductors, and software and computer services) as a primary example of a high-tech industry with a high degree of global interaction and regional concentration. California's computer cluster illustrates very clearly what we mean by high-tech, globalization, and regional concentration:

- Innovation has led to a stream of product developments that have shaped the demand for computers. Computer, electronics, and software firms account for over one fourth of US spending on research and development, as shown in Figure 1.1. Electronics alone accounts for over one-fourth of patents awarded in 1999, as shown in Figure 1.2.
- The industry intensively uses high-quality human capital for its research, design, and management activities. Management, technical, and professional employees comprise almost 30% of electronics employees, more than half of computer manufacturing employees, and 70% of employees in computer software. For all industries, on average, under 15% of employees are in these categories, as shown in Figure 1.3.
- Headquarter activities are highly centralized, both in California and in specific sub-regions such as Silicon Valley. Figure 1.4 shows that even with dispersal of more standardized production, computer manufacturing employment continues to be concentrated in a small number of states.
- The research and design activities have two types of global linkages: (1) the immigration of foreign-born engineers is an important source of high-quality human capital; (2) specialized technical development or research may occur at a number of other locations scattered internationally, as shown in Figure 1.5, even though the core R&D remains centered in the major US locations such as Silicon Valley.
- The production activities are also highly globalized, since they make frequent use of low-cost foreign production centers. This applies to both hardware and software.

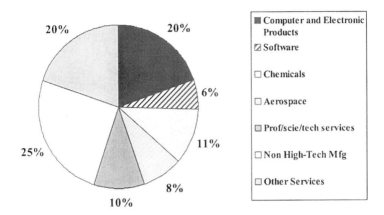

Figure 1-1. Percent of Workers in Selected Managerial, Professional or Technical Professions, 2000. Source: Bureau of Labor Statistics http://stats.bls.gov/oes/home.htm .

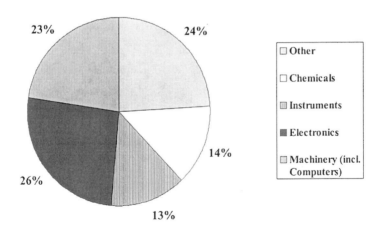

Figure 1-2. Patents Awarded by Industry, 1999. Source: Statistical Abstract of the US, 2001.

Globalization and a High-Tech Economy

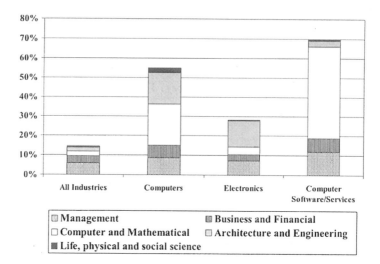

Management	Business and Financial
Computer and Mathematical	Architecture and Engineering
Life, physical and social science	

Figure 1-3. Percent of Workers in Selected Managerial, Professional or Technical Professions, 2000. Source: Bureau of Labor Statistics http://stats.bls.gov/oes/home.htm .

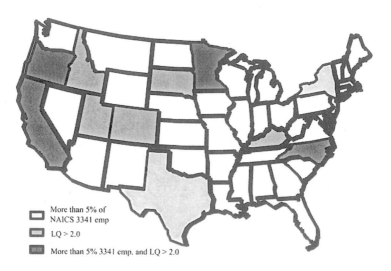

Figure 1-4. Agglomeration of US Computer Manufacturing Employment, 2000. Source: Authors from County Business Patterns data. Lightly shaded areas indicate that the state has more than 5% of the US employment in NAIC 3341 (i.e. computer manufacturing). LQ > 2.0 means that the percentage of the state's employment in computer manufacturing is at least twice the national average. (The location quotient, LQ, is further defined in Chapter 2.)

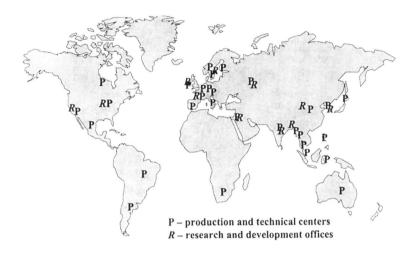

Figure 1-5. Worldwide Locations of Production and R&D Facilities of California Computer Firms. Source: Authors' interviews and web research.

1.2 THE THEORETICAL FRAMEWORK

Our analysis is based on, and is highly consistent with, two recent developments in international trade theory and regional economic theory. First, there is the "new trade theory," see Helpman and Krugman (1985). Second, there is the "new economic geography," as described by Krugman (1991) and Fujita, Krugman, and Venables (1999). Since these views shape how we carry out the analysis, it is useful to understand the key features of this framework. In addition, our analysis provides a unique and important confirmation of the forces that are highlighted in the new international and the new regional economics. We will discuss each of these forces in turn.

1.2.1 Traditional International Trade Theory

The traditional theory of international trade identified "comparative advantage" as the key factor creating international specialization and the associated international trade. Comparative advantage depended in turn on disparate endowments of the factors of production—such as capital, labor, and land—across different regions or countries. This traditional model, for example, explains the location of high-tech industries in countries with large pools of high-skilled labor, that is, human capital. However, there are many observed facts that are in apparent contradiction with these traditional models. Four key conflicts are:

1) International trade in a specific commodity should go in a single direction, originating in the country with the comparative advantage and being shipped to the rest of the world. Empirical evidence that actual exports and imports follow comparative advantage patterns, however, is modest at best.[6] Moreover, for high-tech commodities generally, and specifically for the computer industry in the United States, large commodity trade flows occur in both directions. That is, at the same time the US is exporting large amounts of manufactured computer products, it is also importing large amounts—indeed *larger* amounts—of manufactured computer products.

2) International conglomerates, based on factors such as vertical integration, give rise to distinctive patterns of international, intra-firm trade, which is not readily motivated in the traditional model, at least given the traditional assumption of constant returns to scale production methods. In other words, taking the constant returns to scale assumption literally, we should see large numbers of small firms—described as "backyard capitalism" by Helpman and Krugman (1985)—and not the large integrated enterprises that actually dominate many industries and certainly many high-tech industries.

3) International trade can be defined at a fundamental level as the movement of goods occurring between two nations between which labor cannot easily move. This is accurately reflected in the international trade in computer products, where production occurs in foreign countries to take advantage of the low-cost labor they are able to provide. It is contradicted, however, by the significant number of highly-skilled engineers that immigrate from abroad to work at the headquarters of American computer companies. This is not just a semantic distinction that can be solved by defining the imported engineering talent as imported human capital, and hence trying to connect it to the traditional trade framework. The issue raised is more fundamental: why are the skills of these engineers not tapped in the country of origin through development of more R&D centers than is actually seen in reality. Also, there is evidently an impact on the wages of highly skilled workers relative to the wages of

---

[6] See Deardorff (1984) for a discussion of early empirical studies. Bowen et al, (1987) also failed to confirm the theory of competitive advantage, while a more recent paper by Prasch (1995) argues that theories of competitive advantage are inadequate to deal with a world of large international flows of capital. On the other hand, Noussair et al, (1995), develop an experimental approach that demonstrates behavior consistent with the theory of comparative advantage.

the unskilled component of the labor market in the United States, particularly in the California computer industry. The policy issues raised by immigration and effects on labor markets are fundamental, not semantic.

4) The use of low-cost foreign sites for the production of high-tech products is consistent with the traditional model assuming these locations reflect a comparative advantage. It appears, however, that the specific countries chosen for the foreign production of high-tech products may reflect as well the *networking* patterns of the engineers that immigrated from abroad to work in research and design activities at the firm's headquarters, as well as business-to-business networks; neither of these networking patterns are part of the traditional model.

1.2.2 New International Trade Theory

The new international trade theory broadens the assumptions of the traditional models to explain the complex patterns of international trade and organization structure that dominate the world as we actually see it. Two assumptions are critical for the new theory: *economies of scale* and the possibility of *imperfect competition and monopoly power.*[7]

It has long been understood that *economies of scale* can explain patterns of international specialization that cannot be explained on the basis of the traditional scarce factors model. Economies of scale seem particularly relevant to production of many high-tech products, where the fixed costs of creating a site are often high, but the marginal costs associated with large capacity can be very low. To be clear, the traditional role of low-cost factors of production, such as labor, should still play an important role. The foreign location of large scale plants for producing computer industry products clearly shows both factors at work simultaneously.

Imperfect competition arises in good part as a result of the economies of scale. The recent anti-trust case against Microsoft illustrates the possible importance of market power for a high-tech industry, although this factor plays only a secondary role in this book. On the other hand, our discussion below will emphasize how the combination of economies of scale and imperfect competition plays a key role in understanding two widespread and recurring aspects of globalization, namely multinational firms and intra-industry trade.

---

[7] A key reference for the new international trade literature is Helpman and Krugman (1985).

1.2.3   Traditional Economic Geography Theory

Economic geography has long studied city form and industrial location.[8] Both theoretical and empirical work have addressed the process by which firms choose a specific location, the patterns of economic activities among different locations, and the reasons for differential degrees of urban agglomeration across the landscape. The field of study has introduced many concepts that are of importance to understanding the development of a region, but much of the underlying theoretical development and empirical work focuses on location and spatial patterns within a national context, and many of the models and theories are designed around the "ideal" rather than directed towards understanding the actual. Three key contributions are:

1) Early research focused on the factors that influence firm location.[9] Many of the early works studied cost related factors such as labor costs, transportation costs, the location of raw materials and markets (which are taken as given), and agglomerative characteristics of the industry, based on simplifying assumptions such as perfect competition.[10] On the other hand, several of the authors recognized that when more realistic assumptions are added, such as mobility of labor and the behavior of multiple-location firms, it is no longer possible to reach an equilibrium.[11] This is consistent with a high-tech sector where cost and market factors may have differential effects on firms depending on their stage of development and market conditions in different locations.

2) At the macro level, economic geography looks at the way in which economic activity is spread throughout the landscape. Early work on this topic was highly idealized, starting with a blank plain and trying to explain how urban centers might arise, why agglomeration would occur, and where urban centers would emerge throughout the landscape; see Christaller (1933) and Lösch (1938). With the growing reliance of large corporations on a network of production that extends beyond national boundaries, however, analyses that focus primarily on local or intranational flows miss key issues regarding labor force availability and cost, agglomeration economies (especially those based on human capital), and demand variation in the international marketplace.

3) Growth theory has also contributed to understanding how regional variations emerge within a nation. Two broad theoretical directions make up

---

[8] There are excellent summaries of the intellectual history of location theory in Smith (1971) and Richardson (1977).

[9] See Alonso (1975), Isard (1956), Renner (1947), Smith (1971), and Weber (1929).

[10] See discussions in Alonso (1975) and Smith (1971).

[11] See Richardson (1977), discussion on pages 54 through 57, for example.

this field. For *externally generated growth*, or "export base" theory, policy makers are directed to strengthen the export capacity of their regions (be it domestic or international "exports").[12] A drawback is that external linkages are viewed only in the context of trade, ignoring many other ways a firm can interact with the international economy. For *growth factors internal to the economy* (eg. population and capital investment), the focus is on traditional growth models. However, this approach often fails to recognize the flow of factors of production and demand into and out of the region. Richardson (1977), in addressing the issue of mobility, points to an important characteristic of relevance to our work, commenting that "a substantial proportion of inter-regional capital flows takes place within existing corporations. This reflects the importance of multi-plant firms in a modern industrialized economy."[13]

A key distinguishing factor of the high-tech industry we study is its multi-regional nature. High-tech firms are generally multi-establishment and often have multi-national operations that draw inputs from across the globe. The *effects* of the industry are of importance to individual regions, but an understanding of these effects requires a perspective that goes beyond the geographic boundary of the region.

1.2.4 New Economic Geography Theory

As with the new international trade theory, the new economic geography theories apply broader assumptions to explain the complex patterns of industrial location within a specific country. Economies of scale play a key role here as well. However, there is another related factor, namely *network links and externalities*. Network links can explain a variety of locational decisions for high-tech industries, for both production sites and headquarter sites, that may be difficult to explain on the basis of other traditional factors.

This approach is being increasingly applied in both theoretical and empirical works. On the theoretical side, a key reference for the new economic geography literature is Fujita, Krugman and Venables (2001). This theory demonstrates how increasing returns and transportation costs can lead to upstream and downstream linkages and thus sustained agglomeration. On the other hand, the immobility of some resources can lead to centrifugal forces. One of the earliest empirical works addressing trade, industrial sectors and geographic regions in the "new economic geography" context is Naponen, Graham and Markusen (1993). A set of papers edited by Cox (1997) focus

---

[12] See Tiebout (1962), North (1955), and Thompson (1958).
[13] Richardson (1977), page 107.

on what it means to be a "local" area in a world dominated by global economic networks. Work by Saxenian also highlights the role of global linkages in local centers of high-tech economic activity.[14]

1.2.5   Globalization and A High-Tech Sub-National Region: The Case of California

"If all politics is local, then all economics is regional...."
Barry Eichengreen

This literature is key to studying the globalization of California's high-tech industry. Trade issues are usually analyzed at the national level, but evidence indicates that global integration is important for the increasing specialization and agglomeration effects seen at the level of regions within a country. This is consistent with the assertion by Howes and Markusen (1993) that "Any change in trading patterns ... will reverberate more powerfully on local economies than it does on the national economy."[15] This is particularly true of highly open, fast growing, high-tech sectors.

Sources of this agglomeration include: increasing returns to scale, network externalities, depth and mobility of labor markets, spillover effects, and a growing stock of international contacts, among others. However, this agglomeration also leads to higher wages, higher real estate prices, congestion, and generally higher costs of doing business, thus leading to a new level of sustained disparity between regions. The very nature of a nation-state, with its free flow of capital, labor, and goods and a uniform and homogeneous economic space gives rise to incentives of equalization. In other words, increased regional disparities in return-adjusted costs of doing business will spur firm and labor migration.

The tension and interplay between these disparate forces—those of global economic integration and national market integration—determines the equilibrium level of regional differences, regional specialization, and agglomeration.[16] This interaction also determines the extent to which a region can retain a concentration and agglomeration of industries, even after an industry "matures".

Our analysis of trade at the subnational region leads us to examine the extent to which individual regions—normally, individual states in the US context—should form an industrial and trade policy, and promote such poli-

---

[14] See Saxenian and Edulbehram (1998) and Saxenian, Motoyama and Quan (2002).

[15] Howes and Markusen (1993), page 10.

[16] Markusen (1993) argues that in addition to these forces, policies of "leading firms and host governments" are significant in determining the type of specialization and level of competitiveness of individual regions, not always deliberately (see page 286 of the work cited).

cies through trade organizations, state departments of commerce, and the local chambers of commerce. Given the concentration of "new economy" industries, the diversity of the industrial structure, openness to trade and its location, California is ideally suited to evaluate the theoretical literature discussed above.

The new economic geography and the analysis of global linkages of a regional agglomeration of a high-tech industry help in addressing a weak spot in the literature relating to the interaction between the two major economic forces of globalization and national market integration. This has specific relevance for economic policy in the context of external supply shocks. Global economic downturns could impact regions within a nation differentially, which may require significant regional responses and regional resources.

1.3   POLICY ISSUES RAISED BY OUR ANALYSIS

By way of coming attractions, we list here the major policy issues relating to a high-tech economy, globalization, and regional concentration that are addressed in this book. Four key policy issues raised by globalization are:

1) Low-cost foreign production of high-tech products creates a significant loss of manufacturing jobs in the industry's home region. These job losses create pressure for protectionist policies to "keep the jobs at home".

2) The home region and country face a significant trade deficit (imports far exceeding exports) for the high-tech products. The trade deficit is used as a second argument to "keep the jobs at home". Since many industries with high US trade deficits are not significant in California, for example auto production, a simple typology of imports as neutral (with little California presence), competing (where the presence is significant), and benign (where presence is significant but imports of inputs are critical) helps put state concerns in the context of national trade policy.

3) The demand for highly skilled labor—for example, engineers—is very high in the headquarters' region, leading to rising employment levels and rising wage rates for this sector of the work force. An inflow of foreign engineers is one specific response to the rising demand for highly skilled workers. In contrast, as a result of the loss of manufacturing jobs, production worker jobs and wages fall, creating pressure for a policy solution. Policy proposals range from trade and immigration restrictions to job training for the displaced production workers.

4) Because high-tech activity tends to concentrate in a specific region, its labor market, foreign trade, and immigration effects are magnified. It is thus not surprising that the question whether trade benefits or displaces California businesses and workers has been a continuing debate in California political circles.

While there is an extensive literature on the role of trade in the US economy and case studies exist of the impacts of trade on specific industries or locales, the implications of foreign trade for a state's economy are rarely studied. Our research leads us to the conclusion that foreign trade has a significant effect in expanding California's growth industries, based on the export capacity and profitability of these industries and on the state's Pacific Rim location. We also find that foreign trade is contributing to the reshaping of job opportunities within the state, from manufacturing to non-manufacturing, and from blue-collar to white-collar, jobs. Five key issues raised for a regional state economy with a high-tech agglomeration are:

1) Is global trade a significant factor in the growth of the region's economy? If so, does the trade have special impact on certain industrial sectors?
2) Do local firms adjust their operations in order to compete in global markets? If so, are there important changes in the firms' location decisions or their production processes?
3) Does global trade have a strong impact on the structure of the region's labor market? And more specifically:
   a) Does trade change the occupational distribution within industries?
   b) Does trade affect the distribution of wages and employment between production and non-production workers, within and across industries?
4) How are a region's export industries linked to economic conditions in other parts of the world? For example, does a region's location, or the nature of its immigrant population, play an important role in determining the level and direction of trade?
5) Should a regional state government attempt to encourage trade or to deal with the impacts of trade? If so, what are the best policy tools to use for this purpose?

1.4   OUTLINE OF THE REST OF THIS BOOK

Chapter 2 introduces the definitions and data sources applied in our study of high-tech industries and globalization. The statistical information is used to provide an overview of the key issues resulting from the interaction of a high-tech regional economy with the forces of globalization.

Chapter 3 reports on case-study research of the impact of globalization on California-headquartered, high-tech firms. The chapter reports on interviews and related company information, focusing on the changing global networks in hardware, software, and services, and the implications of these changes for high-tech hubs in California and in other locations globally.

Chapter 4 evaluates the impact that foreign outsourcing has on the markets for skilled and unskilled workers in the US and California labor markets. Econometric methods are used to evaluate whether manufacturing jobs lost in a recession to overseas locations represent temporary or permanent losses. The chapter quantifies the role played by globalization in increased inequality between skilled and unskilled workers and in restructuring by US and California firms during a downturn. It also analyzes the contribution of foreign outsourcing to increased value-addition in manufacturing, particularly in the high-tech sectors.

Chapter 5 looks at the interaction of foreign outsourcing and intra-firm trade, the latter referring to trade between a multinational enterprise and its foreign affiliates. Econometric tests show that foreign outsourcing is increasingly taking place through intra-firm trade, and that this is particularly true for high-tech industries.

Chapter 6 studies the global geography of California's exports, specifically those of the high-tech industry. A modified gravity model is employed to estimate the impact on exports of foreign born immigrants resident in the US and California and the spillover effect of transnational business networks. The model also helps identify those regions of the world with which California has a strong trading affinity.

Chapter 7 examines the actual and potential roles of regional policy in a world increasingly dominated by global networks. An overview of state-level programs throughout the US is followed by a more detailed examination of the example of California, particularly in the context of our research. The chapter concludes with a summary of the options available at the state level for responding to the opportunities and challenges of a global economy.

Chapter 8 provides a summary of the book's conclusions.

Chapter 2

Globalization and a High-Tech Economy
A Statistical Overview of the Issues

This chapter provides an overview of globalization and a high-tech economy using the data sets we have assembled. The accompanying Appendix also offers insights into navigating the multiple data sources and definitions available for the study of high-tech industries.

2.1   DEFINITION OF THE HIGH-TECH SECTORS

In Chapter 1, we provided a conceptual definition of *high tech* based on four key features: innovation, skilled labor, regional agglomeration, and the separation of design and production activities. We now link this conceptual definition to the actual codes used for identifying industry affiliation in US data sources. Until 1997, the classification of industries in US government data was based on a system called the Standard Industrial Classification (SIC) codes. Beginning in 1997, the US government changed to a new coding format, called the North American Industry Classification System (NAICS). Virtually all new data are now collected in the NAICS format, and a limited number of historical series have been converted from a SIC to a NAICS basis by government agencies. For the historical series which remain unconverted, we must create translation tables based on data collected on both a SIC and a NAICS basis for an overlapping year (usually 1997). Since the NAICS system is now the primary system for US government data, we state our definitions in terms of NAICS codes. We describe our methods for linking the unconverted historical SIC data to a NAICS basis in the Appendix.

2.1.1　High-Tech Manufacturing

We begin with the high-tech *manufacturing* sector; in computer parlance, the hardware. The left side of Table 2-1 provides the statistical definition for high-tech manufacturing as NAICS code 3341 (Computers and Peripheral Manufacturing) and NAICS code 3344 (Semi-conductor and Related Manufacturing).[1] The right side of Table 2-1 provides a translation to the equivalent SIC codes, which are primarily components of SIC codes 357 and 367 respectively.[2] Table 2-2 provides the official descriptions of the codes.

Defining an industry cluster requires decisions for what is and is not included. NAICS codes 3341 and 3344 provide a relatively narrow definition of high-tech manufacturing, but still cover the larger part of the high-tech electronics industry, while greatly simplifying data problems. Broader definitions of high-tech manufacturing are frequently based on the American Electronics Association (AeA) definition.[3] The industries included in NAICS codes 3341 and 3344, however, represent the largest part of the AeA definition.[4]

Other research studies have adopted definitions specialized to their specific topic. For example, in studies focused on the research activity of an industry, or on the required skills of its primary workers, occupational codes may be used.[5] To illustrate the range of possibilities, one study has tabulated fourteen different definitions found in the literature.[6] Finally, we do not consider bio-tech, because the issues of bio-tech vary significantly from those of high-tech electronics, and would require a book of their own.

---

[1] The NAICS 4-digit code 3341 has only one 5-digit component, namely 33411, so 3341 and 33411 are identical. Similarly, NAICS codes 3344 and 33441 are identical. We refer to them in this book as 3341 and 3444. Both 3341 and 3344 have a number of 6-digit components; see Table 1.

[2] NAICS 3341 excludes part of SIC 3578 (calculating machines) and excludes all of SIC 3579 (such as watch, clock, and pencil manufacturing). NAICS 3344 excludes part of SIC 3679 (motor vehicle electronics) and adds parts of SIC 366 (telephone apparatus) and SIC 382 (measuring instruments).

[3] For useful discussions of alternative definitions of high-tech see Cortright and Mayer (2001) Appendix B and DeVol (1999) page 34.

[4] The AeA definition also covers parts of SIC 365 (consumer electronics), SIC 366 (communications equipment), SIC 382 (industrial electronics), SIC 386 (photonics), SIC 381 (defense electronics), and SIC 384 (photonics).
See http://www.aeanet.org/Publications/IDMK_definition.asp .

[5] For example, Standard Occupational Codes (SOC) produced by the US Bureau of Labor Statistics are used by Wilkerson (2002).

[6] See Sommers and Carlson (2000), Appendix B, "High Tech Industry Definitions".

Table 2-1. NAICS Definition of High-Tech Manufacturing.

NAICS Code	Industry Name		Links to SIC Codes
3341	Computer and Peripheral Manufacturing Eq		
334111	Electronic Computer Manufacturing	= 3571	Electronic Computers
334112	Computer Storage Device Manufacturing	= 3572	Computer Storage Devices
334113	Computer Terminal Manufacturing	= 3575	Computer Terminals
334119	Other Computer Peripheral Equipment Manufacturing	= 3577 +3578	Computer Peripheral Eq NEC Calculating & Accounting Eq (pt)
3344	Semiconductor and Related Manufacturing		
334411	Electron Tube Manufacturing	= 3671	Electron Tubes
334412	Bare Printed Circuit Board Manufacturing	= 3672	Printed Circuit Boards
334413	Semiconductor and Related Device Manufacturing	= 3674	Semiconductors and Related Devices
334414	Electronic Capacitor Manufacturing	= 3675	Electronic Capacitors
334415	Electronic Resistor Manufacturing	= 3676	Electronic Resistors
334416	Electronic Coil, Transformer, etc. Manufacturing	= 3677 +3661 +3825	Electronic Coils, Transformers, Telephone Apparatus (pt) Measuring Instruments (pt)
334417	Electronic Connector Manufacturing	= 3678	Electronic Connectors
334418	Printed Circuit Assembly Manufacturing	= 3679 +3661	Computer Peripheral Eq (pt) Telephone Apparatus (pt)
334419	Other Electronic Component Manufacturing	= 3679	Electronic Components, NEC (pt)

Abbreviations:

Eq = equipment; NEC = not elsewhere classified.

(pt) indicates that only that part of the SIC code is included in the NAICS code.

Source: US Census Bureau, "Bridge Between NAICS and SIC", EC97X-CS3 (June 2000).

Table 2-2. NAICS Code Description of High-Tech Manufacturing Industries.

3341 Computer and Peripheral Equipment Manufacturing This industry comprises establishments primarily engaged in manufacturing and/or assembling electronic computers, such as mainframes, personal computers, workstations, laptops, and computer servers; and computer peripheral equipment, such as storage devices, printers, monitors, input/output devices and terminals. Computers can be analog, digital, or hybrid. Digital computers, the most common type, are devices that do all of the following: (1) store the processing program or programs and the data immediately necessary for the execution of the program; (2) can be freely programmed in accordance with the requirements of the user; (3) perform arithmetical computations specified by the user; and (4) execute, without human intervention, a processing program that requires the computer to modify its execution by logical decision during the processing run. Analog computers are capable of simulating mathematical models and comprise at least analog, control, and programming elements.
3344 Semiconductor and Other Electronic Component Manufacturing This industry comprises establishments primarily engaged in manufacturing semiconductors and other components for electronic applications. Examples of products made by these establishments are capacitors, resistors, microprocessors, bare and loaded printed circuit boards, electron tubes, electronic connectors, and computer modems.
Source: US Bureau of the Census: http://www.census.gov/epcd/www/NAICS.html .

2.1.2   High-Tech Services

High-tech services—in computer parlance the software—include:

- Computer systems design and related services,
- Software publishers,
- On-line information services,
- Data processing services.

The left side of Table 2-3 shows the NAICS codes for these industries, and Table 2-4 provides the official descriptions of these codes. The right side of Table 2-3 translates the NAICS codes into their SIC equivalents, all components of SIC 737 (Computer Programming, Data Processing, and Other Computer Related Services).[7] SIC code 737 represents the core definition of high-tech service industries in many other studies as well.[8]

---

[7] Unlike the SIC 737 definition, the NAICS definition drops software reproduction, which is now considered a manufacturing process (NAICS 334611), and computer leasing and store retailing and repairing computers, now designated as leasing or retailing activities. Wilkerson (2002) adopts these adjustments but also includes NAICS codes 514110 (news syndicates) and 514120 (libraries and archives), which do not fit our concept of high-tech.

[8] Cortright and Mayer (2001) adopt exactly SIC code 737 as their definition. DeVol (1999) also adopts SIC code 737, but adds SIC 781 (motion picture production and services), SIC 871 (engineering and architectural services), and SIC 873 (research and testing services).

Table 2-3. NAICS Definition of High-Tech Services.

NAICS Code	Industry Name		Links to SIC Codes	
541511	Computer Program Services	=	7371	Computer Programming Services
541512	Computer Systems Design Services	=	7373 +7379	Computer Systems Design Other Computer (pt)
541513	Computer Facilities Services	=	7376	Computer Facilities Services
541519	Other Computer Related Services	=	7379	Other Computer Services (pt)
511210	Software Publishers	=	7372	Software Publishers
514191	On-Line Information Services	=	7375	Information Retrieval Services
514210	Data Processing Services	=	7374	Computer Processing and Data Services

(pt) indicates that only that part of the SIC code is counted in the NAICS code.
Source: US Census Bureau, "Bridge Between NAICS and SIC", EC97X-CS3 (June 2000).

Table 2-4. NAICS Code Descriptions of High-Tech Service Industries.

54151 Computer Systems Design and Related Services
(= 541511 + 541512 + 541513 + 541519)
This industry comprises establishments primarily engaged in providing expertise in the field of information technologies through one or more of the following activities: (1) writing, modifying, testing, and supporting software to meet the needs of a particular customer; (2) planning and designing computer systems that integrate computer hardware, software, and communication technologies; (3) on-site management and operation of clients' computer systems and/or data processing facilities; and (4) other professional and technical computer-related advice and services.

511210 Software Publishers
This industry comprises establishments primarily engaged in computer software publishing or publishing and reproduction. Establishments in this industry carry out operations necessary for producing and distributing computer software, such as designing, providing documentation, assisting in installation, and providing support services to software purchasers. These establishments may design, develop, and publish, or publish only.

514191 On-Line Information Services
This US industry comprises Internet access providers, Internet service providers, and similar establishments primarily engaged in providing direct access through telecommunications networks to computer-held information compiled or published by others.

514210 Data Processing Services
This industry comprises establishments primarily engaged in providing electronic data processing services. These establishments may provide complete processing and preparation of reports from data supplied by customers; specialized services, such as automated data entry services; or may make data processing resources available to clients on an hourly or time-sharing basis.

Source: US Bureau of the Census; http://www.census.gov/epcd/www/NAICS.html .

2.2   HIGH-TECH ACTIVITY

We start with *high-tech manufacturing* shipments.[9] Figure 2-1 shows the annual shipments starting in 1987 for the two components of our definition of high-tech manufacturing, computers and peripheral equipment (NAICS 3341) and semiconductors and related manufacturing (NAICS 3344). Figure 2-1 vividly displays the great computer boom of the 1990s for the two categories of high-tech shipments. However, both the beginning and the end of the decade reflect recessions. The recession from the late 1980s through 1991 led to little or no growth in high-tech shipments. The recession at the end of the 1990s led to significant declines in both categories of high-tech manufacturing shipments. Sources of the latter downturn include the end of the dot-com period and the economic recession that followed.

Next we consider the activity level for *high-tech services*, which we measure as annual revenues, since shipments data are not available. Figure 2-2 shows annual high-tech service sector revenues, as well as the total high-tech manufacturing shipments (the sum of the two curves shown in Figure 2-1). The overall activity levels for the manufacturing and service components of the high-tech industry are roughly equal in magnitude. However, high-tech services have been growing at a substantially faster pace. For the boom decade of the 1990s, the average annual (compounded) growth rate of high-tech service revenues is 14.1% annually, whereas the growth rate of high-tech manufacturing shipments is 8.5% annually. In comparison, the nominal growth rate of the US economy during this period, based on gross domestic product (GDP), was 5.5%. Thus, both sectors of the high-tech economy performed very well, but high-tech services was the star performer.

Due to the current unavailability of service sector revenue data after 2000, we cannot determine the impact of the recession of the early 2000s on high-tech service revenues. However, high-tech service employment levels are available through 2002, and they indicate only a limited impact of the recession (see Figure 2-4 below). This contrasts with the sharp decline in shipments of manufactured high-tech goods starting in 2000. One explanation for the limited recessionary impact on high-tech services is that they depend to an important extent on the *installed computer base,* which keeps rising, whereas high-tech manufacturing shipments depend on *the growth rate of the computer base*, which fell sharply during the recession.

---

[9] Sources for all figures in this chapter are provided in the chapter's Appendix.

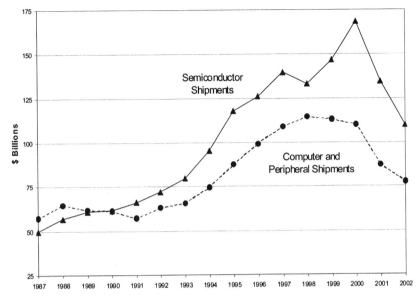

Figure 2-1. Annual US Shipments, High-Tech Manufacturing. High-tech manufacturing covers computers (3341) and semiconductors (3344). NAICS codes in parentheses. Source: Annual Survey of Manufactures, US Census Bureau. See Appendix for details.

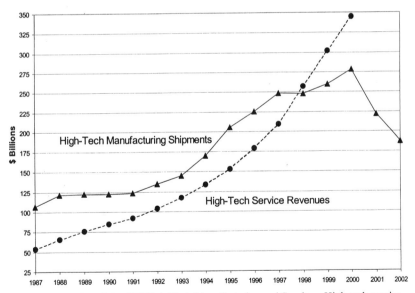

Figure 2-2. Annual Activity, US High-Tech Manufacturing and Services. High-tech service activity is the total revenue for computer design (54151), software publishing (511210), on-line services (514191), and data processing (514210). Data for high-tech services are not currently available beyond 2000. NAICS codes in parentheses. Sources: For services, Service Survey Annual, US Census Bureau; For manufacturing, Figure 2-1. See Appendix for details.

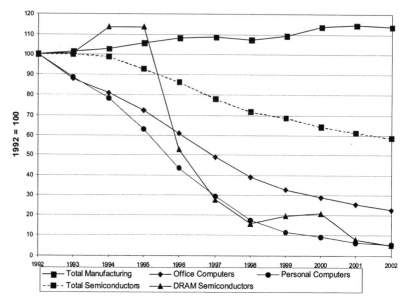

Figure 2-3. US Producer Prices (Quality Adjusted) for High-Tech Manufactured Products. Source: Producer Price Index, US Bureau of Labor Statistics. Indices set for 1992 = 100.

2.3   HIGH-TECH PRICES

Many factors were responsible for the high-tech boom of the 1990s. Here we focus on one factor that is particularly relevant to this book, the rapidly falling prices for high-tech products. Figure 2-3 compares quality-adjusted producer prices for the overall manufacturing sector with several products within the high-tech domain. Overall manufacturing sector prices, shown as the slightly rising line at the top of Figure 2-3, rose at a 1.3% average annual inflation rate over the period. In contrast, prices for all four high-tech products fell dramatically. Personal computer prices had the greatest decline, a 25.4% annual *de*flation rate, and total semiconductor product prices had the smallest decline, a 5.2% annual *de*flation rate. As one explanation for the more rapidly falling computer prices, Aizcorbe, Flamm, and Khurshid (2002) note that semiconductor price declines varied significantly across product types, with the greatest declines for semiconductor products used in computers (such as DRAM chips and microprocessors).

The price indices in Figure 2-3 are all *quality adjusted*, meaning they correspond to products of fixed technical ability throughout the period.[10] Thus,

---

[10] Holdway (2001) discusses how the Bureau of Labor Statistics adjusts computer prices for quality change.

if the *actual market prices* for computer products were constant, the *quality-adjusted prices* would be falling because the quality of the shipped products was rising. This implies that the rising trends in Figures 2-1 and 2-2 for high-tech activity actually understate the rise in *quality-adjusted activity*, since the quality of the products and services is itself rising. On the other hand, the computer shipments data may or may not understate the actual *number* of computers shipped, since this depends on the trend in market prices for the computers sold, whatever their quality. Indeed, to measure the number of units shipped over time would be quite misleading, as the machines sold a decade ago are essentially different products from the machines sold today in terms of their performance.

Producer price indices for high-tech services, comparable to the hardware prices shown in Figure 2-3, have been reported by the Bureau of Labor Statistics only since 1998. These prices show disparate trends: for example, application software prices rose by 4.7% from 1998 to 2002, while computer game software prices fell by 45.1%. The Bureau of Economic Analysis has also reported an implicit price deflator for computer programming, data processing, and related services for a longer period (since 1987). Prices in this aggregate sector have changed very little over this time period, with the 2001 index about 4% above the 1987 level. A chained price index is also available for software (in aggregate, dating back as far as 1959). This index shows a modest decline of 12.9% in software prices between 1990 and 1998, with an increase of 4% between 1998 and 2001. Market domination by a few firms in some key software sectors may be one of the reasons that software prices have fallen less overall than have hardware prices.

2.4   HIGH-TECH EMPLOYMENT AND PRODUCTIVITY

We now look at employment in US high-tech industries. Figure 2-4 shows the number of workers (in thousands) employed in high-tech manufacturing and services industries. High-tech service employment is clearly the growth leader, adding more than 1.5 million jobs between 1987 and 2001, and then losing only 10,000 jobs in the recession through 2002. In contrast, the number of semiconductor manufacturing jobs is approximately constant at the level of 550,000, and more than 100,000 jobs were lost due to the recession starting in 2000. For computer manufacturing, there is a steady job loss, except for a brief uptick during the dot-com boom, such that more than 160,000 jobs were lost between 1987 and 2002. *By 2002, high-tech service sector jobs exceed high-tech manufacturing jobs by a ratio of more than three to one.*

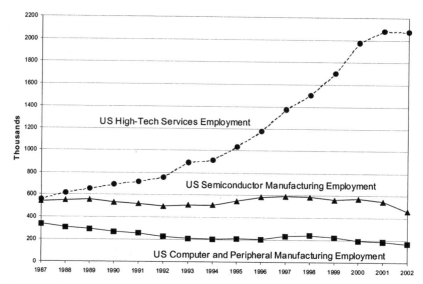

Figure 2-4. US Employment for High-Tech Manufacturing and Services. Definitions of high-tech manufacturing and services as in Figures 2-1 and 2-2. NAICS codes in parentheses. Source: County Business Patterns, US Census Bureau. See Data Appendix for details.

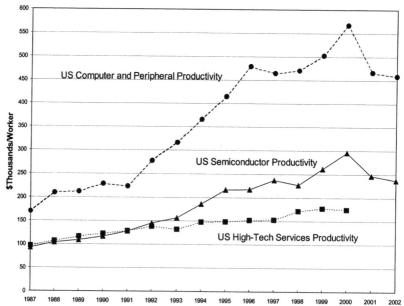

Figure 2-5. US Average Annual Worker Productivity, High-Tech Manufacturing and Services. Average productivity = annual activity level (shipments or revenues)/employment. Productivity data are not currently available for services after 2000. NAICS codes given in parentheses. Sources: See Figures 2-1, 2-2, and 2-4.

Figure 2-5 combines the information we have on high-tech activity and high-tech employment to show the average worker productivity, defined as annual activity/average employment. High-tech service sector productivity is lower than productivity in either high-tech manufacturing sector. This is understandable, since labor is the primary input factor for the service industry, and thus revenue and employment will generally move together. The growing productivity in the manufacturing sectors is driven by three key factors:

1) Technological advances in manufacturing methods allow rising output per worker.
2) More capital-intensive production methods—machines replacing workers—also raise productivity.
3) Material inputs are increasingly imported from foreign sources, thus raising worker productivity.

The third factor is particularly important for the computer industry, and is a likely source of the high growth rate and high level of its productivity (over $450,000 of annual output per worker in 2002).[11] The decline in manufacturing productivity after 2000 is no doubt due to the recession. Productivity data are not yet available for high-tech services after 2000.

2.5 HIGH-TECH INTERNATIONAL TRADE

2.5.1 Trade in High-Tech Manufactured Goods

Figure 2-6 shows trade data for computers and peripherals (NAICS 3341) since 1989. Exports and imports generally grew until the recession starting in 2000. Imports, however, grew faster than exports, so that the net US trade position in computer related manufactured goods switched from a surplus to a deficit in 1993, and that deficit has grown almost every year since, reaching almost $35 billion by 2002.

Figure 2-7 shows the corresponding trade data for semiconductors and related manufactured goods (NAICS 3344). Here too, imports and exports generally grew until 2000. A trade deficit in semiconductors has existed at least since 1989, reaching a peak deficit level of about $35 billion in 1995. Since that time, this trade deficit has moderated, reaching a deficit level of about $10 billion in 2002. Combining NAICS codes 3341 and 3344, the net US trade position in high-tech manufactured electronic goods has had a continuing and significant trade deficit (see Figure 2-9 below).

---

[11] The strong growth in high-tech manufacturing is further discussed in Oliner and Sichel (2002) and Stiroh (2002).

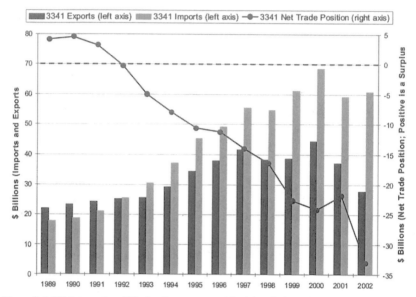

Figure 2-6. US International Trade, Computers and Peripherals (NAICS 3341). The bar graphs shows US exports and imports for NAICS 3341 (left axis). The line graph shows the US trade position in NAICS 3341 (right axis). Source: US International Trade Commission.

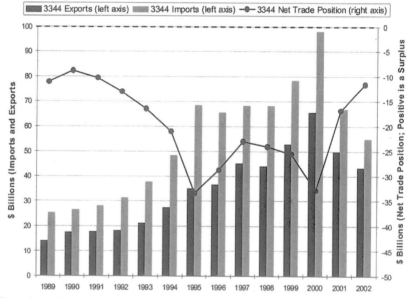

Figure 2-7. US International Trade, Semiconductors and Related Manufacturing (NAICS 3344). The bar graphs shows US exports and imports for NAICS 3344 (left axis). The line graph shows the US trade position in NAICS 3344 (right axis). Source: see Figure 2-6.

Figures 2-6 and 2-7 demonstrate that while the US dominates the world economy for high-tech electronics, the country has systematically imported more of these goods than it has exported. Understanding this phenomenon is a major goal of this book. One key point is that the US high-tech economy has steadily and increasingly used low-cost foreign production centers to manufacture its hardware. The foreign plants are owned either by the foreign affiliates of US parents or by independent contractors. In either case, the benefit is that high-tech hardware costs keep falling, which in turn has led to rising hardware sales (in both the US and abroad). The US economy benefits from this growth in two fundamental ways. First, as we have already seen, activity levels and employment in US high-tech *service* industries have grown steadily and rapidly (recall Figures 2-2 and 2-4), offsetting any declines in high-tech manufacturing. Second, US companies benefit directly from off-shore, high-tech, production in terms of royalties and earnings from foreign direct investment. We now look at these data.

2.5.2   Trade in High-Tech Services

Figure 2-8 shows the major components and total for the US trade balance in high-tech services. High-tech net royalties (receipts minus payments) is the largest component and has been rising steadily. Earnings on high-tech foreign direct investments are the second largest component, but have fallen since 2000 due to the recession-based decline in the profits of the foreign affiliates of US multinational companies (MNCs). High-tech net sales of services by US resident firms is the third and smallest of the components.

Figure 2-9 provides a summary of the overall contribution of high-tech electronic industries to the US trade position, using data already presented in Figures 2-7 and 2-8. High-tech manufactured goods created a continuing trade deficit, reaching over $55 billion in 2000, prior to the recession of the early 2000s. High-tech services, in contrast, created a continuing trade surplus, reaching about $26 billion in 2000. The overall effect is still a trade deficit, but the magnitude is significantly moderated by the trade surplus in services. Furthermore, these data do not incorporate the additional benefits to the US economy resulting from the profits earned by US high-tech firms and the net amount of domestic jobs created.

The US trade deficit in manufactured high-tech goods is particularly intriguing in two respects. First, as shown in Figure 2-9, the US is a net importer of the same manufactured high-tech products for which it dominates the world technology. Second, as shown in Figures 2-6 and 2-7, the US simultaneously imports and exports large quantities of these high-tech products.

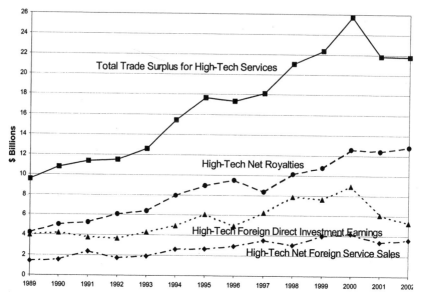

Figure 2-8. US International Trade Balances, High-Tech Services. US international trade in high-tech services consists of the three components shown. Sources: "U.S. International Services" in Survey of Current Business, October 2002, and "U.S. Direct Investment Abroad" in Survey of Current Business, September 2002. See Data Appendix for details.

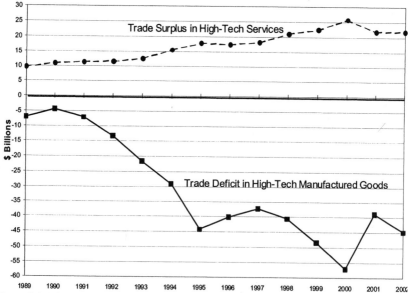

Figure 2-9. US International Trade Balances, High-Tech Manufacturing and Services. The trade balance in high-tech manufactured goods is the sum of the trade balances for NAICS codes 3341 and 3344, from Figures 2-6 and 2-7 respectively. The trade balance in high-tech services is from Figure 2-8. Sources: See Figures 2-6, 2-7, and 2-8.

2.5.3   Imported Intermediate Inputs

Imported intermediate inputs provide an explanation for the two-way pattern of US international trade in manufactured computer products. Inputs are imported from abroad to take advantage of the low costs available in foreign production centers. The imported parts are assembled into complete products, and then either sold to the US market or exported abroad. For some firms, such as Dell Computers, their products consist primarily of imported inputs, with only the final assembly, possibly only the final shipping, taking place in the US.

The importance of imported inputs to industries can be most readily demonstrated by computing *import shares*, defined as:

$$\text{Import Share} = \frac{\text{Imports}}{\text{Shipments} + \text{Imports} - \text{Exports}}$$

The denominator of the ratio is the total supply of goods available to satisfy domestic demand. The import share is thus the percentage of total domestic demand satisfied by imports.

Table 2-5 shows the top 12, 4-digit, NAICS industries in the order of their import share ranking for 2000, as well as the data used to compute the import shares. The list represents the "usual suspects" for industries for which the US economy has large amounts of imports: consumer electronics, footwear and clothing (including leather), and motor vehicles.

The high-tech industries, NAICS 3341 and 3344, also rank among these top industries for import shares. In both cases, approximately 50% of the domestic market is served through imported products. Furthermore, the two industries rank 2nd and 3rd among all 4-digit NAICS industries for imports (motor vehicles are number 1) and 1st and 3rd for exports. Finally, the two industries rank 4th and 11th among all the 4-digit NAICS industries for shipment volume. Overall, Table 2-5 provides a strong demonstration of the key role that global trading plays for our high-tech industries.

An industry can have a high import share for two different reasons:

1) Large amounts of finished products are directly imported into the US for sale to consumers.
2) Large amount of imported inputs are used in the US for producing the final goods.

Both forces may be active in an industry, and import shares by themselves cannot identify which is the primary factor for a given industry. For example, motor vehicle imports into the US include both cars fully assembled abroad and components that will be assembled into finished cars in the US.

Industries with *both high import shares and high export amounts*, however, provide a signal that a significant part of the imports are inputs to the production process, since it would not make sense for an industry to import and to export the same finished goods.[12] It is noteworthy in this regard that NAICS 3341 and 3344 are the only industries in Table 2-5 for which their exports are as close as or closer to the top ranking (i.e. # 1) than are their imports. Chapter 4 provides a more complete discussion of the key role performed by imported inputs in the US high-tech industries.

Table 2-5. US Import Shares and Related Data for the Top 12, 4-Digit, NAICS Code Industries, 2000.

Code	Industry Name	Import Share	$ Billions			Rank			
			Im-ports	Ex-ports	Ship-ments	Imp. Sh.	Im-ports	Ex-ports	Ship-ments
3343	Audio & video	84.8%	28.7	4.2	9.3	1	10	38	77
3162	Footwear	81.0%	14.5	0.4	3.8	2	18	80	84
3169	Leather & allied	72.4%	4.9	0.8	2.7	3	40	73	86
3152	Cut & sew apparel	58.3%	58.4	6.3	48.0	4	4	27	31
3314	Nonferrous metal	57.3%	18.0	10.4	23.7	5	15	18	57
3159	Apparel accessories	52.9%	3.6	1.4	4.6	6	50	57	83
3341	**Computer & peripheral**	51.0%	68.5	44.3	110.0	7	3	3	11
3161	Leather & hide tanning	50.4%	2.0	1.2	3.2	8	64	64	85
3344	**Semiconductors**	48.7%	98.1	65.2	168.5	9	2	1	4
3399	Other miscellaneous	48.7%	48.5	9.8	61.0	10	5	20	20
3333	Commercial Machinery	38.6%	12.4	8.4	28.1	11	22	24	48
3361	Motor Vehicles	37.4%	129.4	23.0	239.4	12	1	8	1

Import shares measure the percentage of a product's final US demand that is served by imports. Final demand is measured as Shipments + Imports − Exports. The import share formula is: Import share = Imports/(Shipments + Imports − Exports). Data sources: for shipments, see Figure 2-1; for imports and exports, see Figure 2-6.

---

[12] The auto industry represents an exception in this regard, since US manufacturers are exporting American-made cars at the same time that foreign manufacturers are importing foreign-made cars into the US.

2.6   A REGIONAL HIGH-TECH ECONOMY: THE CALIFORNIA CASE STUDY

California's Gross Domestic Product now represents just under 14% of the overall US economy, making the state the world's fifth largest economy by this measure. Beyond size, a primary feature of the California economy is the intensity of its high-tech industries, as illustrated by Silicon Valley in Northern California. In this section, we look at the relevant statistical features of California as our case study. The discussion documents why California is an excellent example of a high-tech, regional economy and demonstrates the essential role of globalization for such an economy.[13]

2.6.1   The Key Role of California's High-Tech Industries

We use a statistical measure known as the *location quotient* to demonstrate the importance of high-tech industries for California. A location quotient equals the ratio between the share one industry represents as a percentage of the regional economy (California) and the share that the same industry represents as a percentage of the total US economy:

$$\text{Location Quotient} = \frac{\text{Industry State Share}}{\text{Industry US Share}} = \frac{(\text{State industry/State aggregate})}{(\text{US industry/US aggregate})}.$$

If an industry's regional share equals its national share, then the location quotient is 1.0. A location quotient greater than 1.0 indicates the industry is more important in the regional economy than in the national economy, and vice versa if a location quotient is less than 1.0. Location quotients can be computed for any industry activity, such as employment and sales.

Table 2-6 shows California's employment location quotients for 2-digit NAICS industries, ranked from highest to lowest. The two highest are Professional, Scientific, and Technical Services (NAICS 54) and Information Services (NAICS 51). These two 2-digit NAICS codes include all our high-tech service industries. Other high-ranking sectors, such as real estate, arts, and forestry are also understandable. The location quotient for California's manufacturing sector (NAICS 31), is below average, at only 94% of the US level. We next look at 3- and 4-digit California location quotients.

---

[13] In principle, trade linkages between California and the other states of the US could also be studied, and these would certainly be of interest. Unfortunately, virtually no data are available for inter-state trade by industry. Indeed, the need to collect customs duties is the primary reason that *international* trade data are readily available.

Table 2-6. California Employment Location Quotients for 2-Digit NAICS Code Industries.

2-Digit NAICS Code	Industry	Employment Location Quotient
54	Professional, scientific & technical	1.38
51	Information services	1.32
53	Real estate	1.27
71	Arts, entertainment & recreation	1.26
11	Forestry, fishing, agriculture	1.17
42	Wholesale trade	1.17
56	Waste Management	1.13
23	Construction	1.02
72	Accommodation & food services	1.00
48	Transportation	0.98
31	Manufacturing	0.94
81	Other services	0.93
52	Finance & insurance	0.92
61	Educational services	0.92
55	Enterprise Management	0.90
44	Retail trade	0.89
62	Health care	0.83
22	Utilities	0.69
21	Mining	0.38

Employment location quotients equal each industry's percentage of total California employment relative to that industry's percentage of total US employment. Data source: *County Business Patterns* 2000, US Census Bureau.

2.6.1.1   California Location Quotients for High-Tech Manufacturing

Table 2-7 shows California's location quotients for 3- and 4-digit NAICS manufacturing industries for 2000. Part A of the table shows location quotients for the top ten, 3-digit, NAICS manufacturing codes. Industries are listed in the order of their location quotient based on shipments. Computers and Electronics (NAICS 334) ranks first by a large margin; our two 4-digit high-tech industries, Computers (NAICS 3341) and Semiconductors (NAICS 3344), are both included within this 3-digit category. The only serious competitor is Apparel Manufacturing (NAICS 315). While this may surprise some, Southern California benefits from a major apparel industry, reflecting the state's proximity to Far Eastern and Mexican suppliers and labor.

Table 2-6 showed that California is less intensive in manufacturing than is the overall US economy. This is confirmed in part A of Table 2-7, which shows location quotients less than 1.0 for industries beyond the top 6. Part A of Table 2-7 also shows California location quotients based on other industry measures, such as material costs and value added. Computers and Electronics is either first or second (with Apparel manufacturing again the only competitor) for each of the alternative measures.

Table 2-7. California Location Quotients for 3-Digit and 4-Digit NAICS Code Manufacturing Industries

Part A Top 10 3-Digit NAICS

NAICS	Industry Name	Shipment Value	Material Costs	Value Added	All Employees		Production Workers	
					Number	Payroll	Number	Payroll
334	**Computer & Electronics**	**2.67**	**2.93**	**2.41**	**2.14**	**2.32**	**2.07**	**2.52**
315	Apparel Manufacturing	2.27	2.58	1.96	2.20	1.97	2.40	2.25
339	Miscellaneous Manufacturing	1.65	1.57	1.59	1.44	1.47	1.51	1.58
312	Beverage & Tobacco	1.07	1.73	0.75	1.53	1.38	1.56	1.53
324	Petroleum & Coal	1.04	0.95	1.63	1.11	1.20	1.15	1.45
337	Furniture & related	1.03	1.19	0.88	1.06	0.95	1.14	1.12
323	Printing & related	0.94	1.01	0.85	0.96	0.91	1.03	1.11
332	Fabricated metal	0.93	0.95	0.88	0.97	0.94	1.08	1.14
311	Food Manufacturing	0.92	0.99	0.85	0.94	0.92	1.01	1.09
327	Nonmetallic Mineral	0.87	1.07	0.70	0.79	0.78	0.86	0.94

Part B Top 10 4-Digit NAICS

NAICS	Industry Name	Shipment Value	Material Costs	Value Added	All Employees		Production Workers	
					Number	Payroll	Number	Payroll
3341	**Computer & Peripherals**	**3.08**	**3.20**	**3.02**	**2.08**	**2.21**	**1.85**	**2.34**
3346	Magnetic & Optical Media	2.85	3.26	2.48	2.88	2.75	2.81	2.85
3332	Industrial machinery	2.57	2.93	2.29	1.59	1.63	1.55	1.71
3152	Cut & sew apparel	2.55	2.91	2.18	2.51	2.28	2.76	2.63
3344	**Semiconductor Manufacturing**	**2.50**	**3.04**	**2.12**	**2.28**	**2.54**	**2.34**	**2.86**
3345	Instrument Manufacturing	2.14	2.39	1.89	1.76	1.91	1.56	1.97
3391	Medical equipment & Supplies	2.12	1.98	2.00	1.71	1.81	1.72	1.88
3114	Fruit & vegetable	1.87	2.18	1.57	1.99	1.88	2.17	2.34
3159	Apparel accessories	1.65	1.65	1.59	1.93	1.73	2.18	2.12
3362	Motor vehicle bodies	1.55	1.85	1.18	0.77	0.71	0.87	0.85

Location quotients for any industry activity equal the percentage that the industry activity represents of the corresponding California manufacturing total, relative to that industry's percentage of the corresponding US manufacturing total. Data source: *Annual Survey of Manufactures*, 2000, US Census Bureau.

Part B of Table 2-7 shows location quotients for the top-10, 4-digit, NAICS codes (again based on shipment values). Our two high-tech sectors, Computers (NAICS 3341) and Semiconductors (NAICS 3344) rank first and fifth among the 4-digit manufacturing location quotients for California. The quotients indicate that these California industries exceed the nation's intensity by factors of 3.08 and 2.50 times respectively. Other top-10 industries include two other technical sectors (NAICS 3345 and 3346).[14]

The NAICS 3341 and 3344 location quotients for material costs exceed the corresponding quotients for shipments, indicating an extensive use of material inputs. This is another indicator of the key role of outsourcing for high-tech production. The location quotients for the number of employees, in contrast, are lower than the quotients for shipments and material costs, indicating that material inputs and capital intensive techniques are substituted for labor. This is consistent with the falling trend for high-tech manufacturing employment and the rising trend for average high-tech manufacturing labor productivity, shown earlier in Figures 2-4 and 2-5.

2.6.1.2   California Location Quotients for High-Tech Services

Table 2-8 shows California's employment location quotients for the NAICS 51 and 54 service industries for the year 2000. The top two service industries are motion pictures and recording, clearly key industries for California. Two high-tech industries, Software Publishers and Information Services come next, while Computer Systems Design and Data Processing services follow on the list. The relatively low quotient for Data Processing reflects the efficiency of distributing processing centers across the country.

2.6.1.3   Location Quotients for Silicon Valley California

Location quotients can be computed for areas smaller than a state. Figure 2-10 shows the largest employment location quotients for Santa Clara County, the home of Silicon Valley. For high-tech sectors such as computers, semiconductors, and software publishers, the location quotients are above 10.0. Indeed, all of the industries with location quotients greater than 4.0 fall within the high-tech cluster. Five of these are manufacturing sectors, accounting for over 125,000 jobs (61% of the county's manufacturing jobs). Four of the remaining five are services sectors, accounting for close to 80,000 additional jobs. Wholesale trade in electronic goods (27,000 jobs) is also among the top ten Santa Clara County location quotients.

---

[14] It is also not surprising that fruit and vegetable manufacturing (NAICS 3114) and medical equipment and supplies (NAICS 3391) appear as intensive industries for California. While some of the other industries may seem perplexing in terms of their 4-digit descriptions, there must be a specialized industry at the 5-digit level.

Table 2-8. California Employment Location Quotients for 4-Digit NAICS Service Codes

4-Digit Components of NAICS 51 and 54		Employment Location
NAICS	Industry Name	Quotient
5121	Motion picture & video industries	2.90
5122	Sound recording industries	2.76
5112	**Software publishers**	**2.28**
5141	**Information services**	**2.17**
5412	Accounting, tax prep, bookkeeping, etc.	2.16
5417	Scientific R&D services	1.72
5414	Specialized design services	1.38
5415	**Computer systems design**	**1.37**
5418	Advertising & related services	1.26
5413	Architectural, engineering & related	1.17
5416	Management, sci & tech consulting	1.10
5411	Legal services	1.07
5142	**Data processing services**	**1.03**
5132	Cable networks & programs	1.02
5419	Other professional, scientific, technical	0.99
5131	Radio & television broadcasting	0.99
5133	Telecommunications	0.93
5111	Database, Newspaper, etc.	0.90

Data source: *County Business Patterns*, 2000, US Census Bureau.

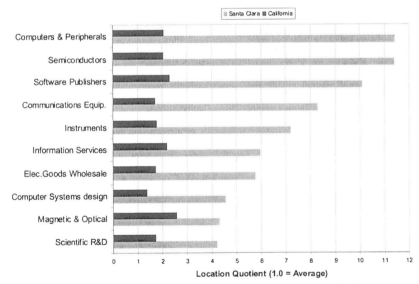

Figure 2-10. Employment Location Quotients for 4-Digit NAICS Codes With the Largest Employment in Santa Clara Country. Data source: *County Business Patterns*, 2000, US Census Bureau.

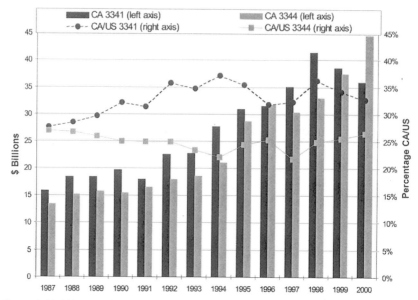

Figure 2-11. Shipments for California High-Tech Manufacturing. High-tech manufacturing covers computers (3341) and semiconductors (3344). NAICS codes in parentheses. Source: Annual Survey of Manufactures, US Census Bureau. See Appendix for details.

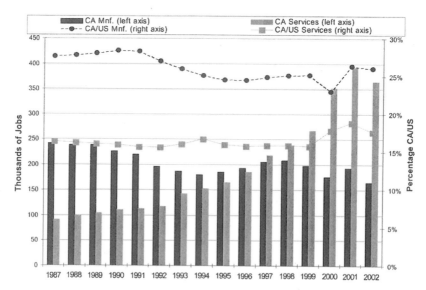

Figure 2-12. California Employment in High-Tech Industries. For definitions of sectors, see Figure 2-4. Source: County Business Patterns, US Census Bureau. See Appendix for details.

2.6.2   Trends in California's High-Tech Industries: Shipments, Employment, and Productivity

Figure 2-11 provides a time series view of shipments in California's two high-tech manufacturing industries.[15] The bar graphs show that the dollar volume of high-tech shipments (measured on the left axis) have been steadily growing over the period, although computer shipments (NAICS 3341) actually decline slightly after 1998. The line graphs (measured on the right axis) show that California's share of US high-tech shipments has been relatively stabile, with some upward gain in California's share of computer shipments. Unfortunately, high-tech service revenue data, comparable to that shown in Figure 2-2 for the US, are not available on a state basis.

Figure 2-12 provides a time series view of employment in California's high-tech industries. The bar graphs (measured as thousands of jobs on the left axis) show that California employment in high-tech manufacturing has been generally falling, while high-tech services employment has been rapidly rising. The line graphs (measured on the right axis) show that California's share of US high-tech manufacturing employment was also falling, at least through 2000. California's share of US high-tech service employment has been steady, but it appears to have moved higher since 2000. Overall, employment in California's high-tech industries generally parallels the US trends.

Figure 2-13 provides views of average labor productivity in California's high-tech manufacturing industries. The bar graphs show that average labor productivity (measured on the left axis) has been steadily and rapidly rising in both industries, similar to the US data shown earlier in Figure 2-5. The line graphs (measured on the right axis) show that, in fact, growth in productivity in both California industries has significantly outpaced the comparable growth for the US, such that, by 2000, California computer productivity (NAICS 3341) is 40% higher than US computer productivity and California semiconductor productivity (NAICS 3344) is almost 20% higher than the US equivalent. As discussed further by Daly (2002) and Wilson (2002), California productivity has significantly exceeded the US value for a wide range of industries in recent years, with California's high-tech industries leading the way. One intriguing explanation is a "Silicon Valley Effect", based on the unique firm structures, workplace practices, and intra-firm spillover effects that characterize the California high-tech industries, in line with the research of Saxenian (1994).

---

[15] The time series data shown in Figures 11, 12, and 13, all end in the year 2000 due to the current unavailability of California data beyond that year.

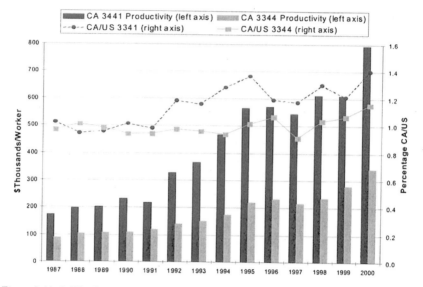

Figure 2-13. California Average Worker Productivity, High-Tech Manufacturing. Average productivity = annual shipments/employment. Sources: See Figure 2-11 for manufacturing shipments; See Figure 2-12 for employment data.

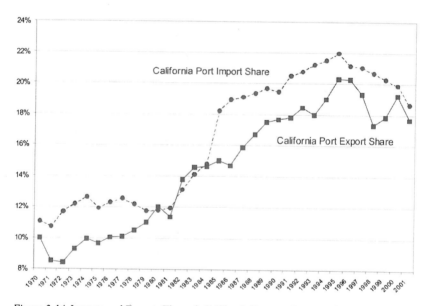

Figure 2-14. Imports and Exports Through California Ports, as Percent of US Total. Source: Calculated by the authors from the following data: California data are from the California Department of Finance; US data are from the Bureau of Economic Analysis, International Transactions Account Data.

2.6.3   International Trade in California's High-Tech Industries

Three datasets link California's economy with international trade:

1) Import and export activity through California ports.
2) California exports (but not imports).
3) US trade patterns applied to California industries.

2.6.3.1   Activity Through California Ports
    California has three major ports, San Francisco, San Pedro Bay (Los Angeles and Long Beach), and San Diego. The San Pedro port is the largest in the US and the third largest in the world. As shown in Figure 2-14, the California port share of total US trade is now about 20%. This amount has about doubled over the last thirty years, due to (a) increasing trade with Asian countries and (b) growth in California and other western-state economies. Exports through California custom districts, however, do not necessarily originate in California, and imports through California ports need not be destined for California locations. Therefore, further data sources are needed to fully determine how trade interacts with California's high-tech economy.

2.6.3.2   California Exports
    The US Census Bureau in affiliation with the Massachusetts Institute for Social and Economic Research (MISER) now estimates state exports based on information provided by shippers.[16] Import data on a state basis, however, are not available due to the difficulty of determining the destination state for US imports. State level trade data for services are also not available.
    Table 2-9 shows California exports for the most important 3-digit NAICS code industries in 2000. Computers and electronic products (NAICS 334) is the most important industry. Exports of this industry represented more than half of the state's total exports, and were almost five times the value of California's next largest merchandise export industry. Of the next six export industries, four also have significant high-tech components or linkages (machinery, transportation equipment, miscellaneous manufactured, and electrical equipment). [17] NAICS 334 exports from California also represent more than 38% of total US exports for this industry, a far higher ratio than any other California industry.

---

[16] Information on this data source and access to the data can be found at
    http://www.ita.doc.gov/td/industry/otea/. More detailed data are available directly from the MISER source, described at http://www.misertrade.org/ . The Miser series fills in gaps in the Census Bureau data using an imputation algorithm.
[17] California transportation equipment is mainly aerospace, while it is motor vehicles for the rest of the US.

Table 2-9. California Exports, 3-Digit NAICS Codes with Largest State Employment, 2000.

NAICS	Industry	California Exports $ Millions	% of State Total	US Exports $ Millions	% of US Total	California % of US
334	**Computers & Electronics**	61,447	51.4%	161,449	25.1%	38.1%
333	Machinery, Not Electrical	13,744	11.5%	85,038	13.2%	16.2%
336	Transportation Equipment	8,158	6.8%	121,701	18.9%	6.7%
325	Chemicals	4,775	4.0%	77,649	12.0%	6.1%
339	Miscellaneous Manufactured	4,107	3.4%	19,328	3.0%	21.2%
311	Food & Kindred	3,434	2.9%	24,966	3.9%	13.8%
335	Electrical Appliances	3,968	3.3%	25,401	3.9%	15.6%
	Total Exports	119,640	100%	644,442	100.0%	18.6%

Source: California Trade and Commerce Agency, MISER data.

Table 2-10. US Trade for 4-Digit NAICS Codes with Largest State Employment, 2000.

NAICS	Sector	% of Total US Manufacturing for Each Category		
		US Imports	US Exports	US Employment
3344	**Semiconductors**	9.4%	10.1%	3.5%
3364	Aerospace product & parts	2.6%	8.0%	2.7%
3341	**Computers & peripherals**	6.6%	6.9%	1.2%
3363	Motor vehicle parts	4.6%	6.6%	4.9%
3345	Navigational & control instruments	2.1%	4.2%	2.8%
3339	Other general purpose machinery	1.9%	3.6%	2.6%
3342	Communications equipment	3.0%	2.9%	1.6%
3254	Pharmaceutical & medicine	2.8%	2.4%	1.4%
3332	Industrial machinery	1.1%	2.1%	1.4%
3329	Other fabricated metal products	1.4%	2.0%	1.1%

Sources: Employment data, *County Business Patterns*, US Census Bureau; Trade data, "Data Web" of the US International Trade Commission, http://dataweb.usitc.gov/ .

2.6.3.3   Trade Impacts for Key California Industries

While data on California imports are not available, imports do play an important role in California industries. Table 2-10 provides a means of measuring the import (and export) intensity of California's high-tech economy. The table shows the ten NAICS 4-digit manufacturing industries having the largest employment in California, in the order of the number of employees. The US import, export, and employment intensities of these industries are computed as the percentage that each value represents of the corresponding US manufacturing total for that category. Industries can be judged as trade intensive when their import and/or export percentage significantly exceeds their employment percentage. Indeed, this condition holds for one or the other of the trade categories for all the industries in Table 2-10. This itself demonstrates the critical importance of trade for the California economy. Furthermore, our two high-tech manufacturing industries, NAICS 3341 and 3341 rank 1st and 2nd for their import percentages and 1st and 3rd for their export percentages. The only competitor is Aerospace products (NAICS 3364), another industry dealing in high-tech products.

Table 2-11 provides one final method for illustrating the relative significance of trade flows for high-tech manufacturing sectors with 25,000 or more employees in California. We compute US imports as a share of domestic consumption and exports as a share of domestic production for these sectors. The table then allocates the sectors into "quadrants" based on their trade shares relative to the average for all US manufacturing industries. Quadrant I industries have above average shares for both exports and imports. Quadrant III industries have below average shares for both exports and imports. Quadrant II and IV industries have above average trade shares for only imports or exports respectively. The table shows that virtually all of California's significant manufacturing sectors lie in either Quadrant I or III. As mentioned earlier in Chapter 1, this conflicts with classical trade theory, which predicts that most industries become specialized in either importing or exporting, but not both.

The key result of Table 2-11 is that all of California's high-tech sectors, and the related aerospace sector, fall into Quadrant I, with above average shares for exports and imports. In other words, all of California's high-tech sectors are also high-trade sectors. The case studies presented in Chapter 3 below illustrate how global integration of high-tech companies may involve both high levels of foreign sales and high levels of international imports.

Table 2-11. Trade Flows Relative to Output and Consumption, California Manufacturing Sectors with 25,000 or More Employees.

	Below Average Imports/Consumption	Above Average Imports/Consumption
Above Average Exports/Shipments	*Quadrant IV* 3391 Medical Eq. & Supplies	*Quadrant I* 3329 Fabricated Metal Products 3332 Industrial Machinery 3339 General Purpose Machines 3341 Computer & Peripheral Eq 3342 Communications Eq 3344 Semiconductors 3345 Instruments Mfg 3363 Motor Vehicle Parts 3364 Aerospace Product & Parts 3399 Other Miscellaneous Mfg
Below Average Exports/Shipments	*Quadrant III* 3114 Fruit & Veg. Preserving 3118 Bakeries and Tortilla Mfg. 3121 Beverages 3219 Other Wood Products 3222 Converted Paper Products 3231 Printing & Related Support 3254 Pharmaceutical Mfg. 3261 Plastics 3323 Structural Metals 3327 Machine Shops 3371 Furniture	*Quadrant II* 3152 Cut & Sew Apparel

Source: Computed by Authors from US International Trade Commission and *Annual Survey of Manufacturers* data.

2.7   SUMMARY AND CONCLUSIONS

This chapter has studied the economics of a high-tech economy. High-tech manufacturing is defined to include computers and semiconductors (NAICS codes 3341 and 3344 respectively). High-tech services include computer system design, software publishing, on-line information systems, and data process services (NAICS codes 54151, 511210, 514191, and 514210 respectively). At the US level, the key findings are:

1) High-tech manufacturing shipments and service revenues rose rapidly during the 1990s, with service revenues surpassing manufacturing shipments starting in 1998. High-tech manufacturing shipments collapsed after 2000, following the end of the 1990s boom.

2) High-tech, quality-adjusted, hardware prices fell dramatically during the 1990s. These price declines were a key source of the high-tech boom.

3) High-tech manufacturing employment has steadily declined in the US, with more than 150,000 jobs lost between 1990 and 2002. In contrast, high-tech service industry employment rose dramatically, with almost 1.5 million jobs added in the same period. By 2002, the ratio of high-tech service jobs to high-tech manufacturing jobs exceeded three to one.

4) Average labor productivity in US high-tech manufacturing is high and growing, reflecting technological advances, capital intensive production methods, and increased outsourcing of component manufacturing.

5) The US has run a systematic trade deficit in high-tech manufactured goods, with the deficit exceeding $50 billion in 2000. At the same time, there has been a systematic US trade surplus in high-tech services, covering royalties, foreign direct investment earnings, and direct foreign service sales. The surplus for services was about $26 billion in 2000.

6) About 50% of the US high-tech hardware market is supplied by imported goods. An important part of these imports are inputs into the production and assembly of high-tech products within the US. Access to low-cost foreign component production thus provides a direct link to falling hardware prices and the industry's growth.

This chapter used California as a case study for the analysis of a high-tech *regional* economy. At the California state level, the key findings are:

1) Location quotients confirm the importance of high-tech manufacturing and services for California's economy. The coefficients also indicate how much these industries rely on material inputs, consistent with the key role of component outsourcing for the production process.

2) Employment trends in California's high-tech industries generally parallel the US trends, with a continuing decline in manufacturing employment and a strong expansion in service employment.

3) The level and growth of productivity in California's high-tech industries exceeds even that of the comparable US industries. One intriguing explanation is the possibility of a "Silicon Valley Effect", based on the unique workplace and industry culture that appears there.

4) Trade through California's port has expanded rapidly during the last 30 years, and this activity is an important part of the California economy. Data on California exports indicate a large volume of international exports for the California economy generally. Furthermore, high-tech exports represent over 50% of California's total exports and over 38% of total US high-tech exports. Unfortunately, state level trade data are not available for imports or for services.

2.8   APPENDIX: DATA DESCRIPTION

This appendix describes the data sources and the techniques we used in compiling the tables and figures of this chapter. The discussion is organized by type of data and the applicable tables and figures. Unless otherwise indicated, all publications are from the US Bureau of the Census.

2.8.1   Manufacturing Sector Data, *Annual Survey of Manufactures* (ASM) (Table 2-7, Figures 2-1 and 2-11)

The *Annual Survey of Manufactures* measures shipments, employment, and related variables for manufacturing industries for the US and individual states, to the 6-digit NAICS industry level. Table 2-7 provides a good example of these data from the year 2000 report. Three points are relevant to the shipments data shown in Figures 2-1 and 2-11:

1) ASM shipments are measured on an "industry" basis, meaning that all shipments from a specific location (such as a plant) are associated with the primary industry of that location, even if some of the shipped goods might be properly associated with another industry. The ASM shipments data also introduce the possibility of double-counting, since goods might be reshipped through different stages of production, but still counted under the code of the primary industry. As an alternative, the US Census Bureau also tracks shipments on a "product" basis, but these data lack the continuity and comparability with employment provided by the ASM. In any case, the same basic trends are apparent in both data sets.

2) ASM data from 1997 forward are provided in NAICS coding. Data prior to 1997 were provided in SIC coding. Table 2-1 provides the links from the SIC to the NAICS codes that were used to restate the historic data on a NAICS basis. In most cases, 100% of a SIC category is linked to a NAICS category. In cases in which only a part of a SIC category is linked to a NAICS code, shown as "pt" in Table 2-1, the following percentages, derived from US Census Bureau, "Bridge Between NAICS and SIC", EC97X-CS3 (June 2000), were applied to the SIC shipments data for 1997 and previous years:

93.00% of SIC 3578 is applied to NAICS 334119
0.022% of SIC 3661 is applied to NAICS 334416
0.175% of SIC 3825 is applied to NAICS 334416
3.43% of SIC 3661 is applied to NAICS 334418
63.44% of SIC 3679 is applied to NAICS 334418
27.09% of SIC 3679 is applied to NAICS 334410

3) The most recent ASM data year currently available is 2000. However, the Census Bureau also carries out a monthly survey of "Manufacturing Shipments, Inventories, and Orders", referred to as the "M3 series". We have used the M3 series to extend the ASM data to 2001 and 2002. In cases where the M3 data were more aggregated than the ASM data, we applied the growth rate of the M3 aggregate to each of the ASM components under that aggregate. Further information on the M3 series is available at http://www.census.gov/mcd/.

2.8.2 Employment Data, *County Business Patterns* (CBP) (Tables 2-6, 2-8, and 2-10; Figures 2-4, 2-10, and 2-12)

The US Census Bureau and the US Bureau of Labor Statistics collect employment data using both individual surveys and enterprise reports. A variety of statistical releases are then provided. The CBP reports from the Census Bureau are particularly useful for our purposes since the data are tabulated by detailed NAICS codes (since 1997) and by states, and they cover both manufacturing and service sectors.

The time series data in Figures 2-4 and 2-12, however, raise two further issues. First, only CBP data from 1997 forward are provided in NAICS coding. Data prior to 1997 were provided in SIC coding. The issue of converting the historical SIC data to a NAICS basis is exactly the same as described above for data from the Annual Survey of Manufactures. Tables 2-1 and 2-3 in the main text provide the links from the SIC to the NAICS codes that were used to restate the historic data to a NAICS basis for the manufacturing employment and service employment data respectively. In cases in which only a part of a SIC category is linked to a NAICS code, shown as "pt" in the tables, the following percentages were applied to the SIC employment data for 1997 and previous years.[18]

87.43% of SIC 3578 is applied to NAICS 334119
.057% of SIC 3661 is applied to NAICS 334416
.299% of SIC 3825 is applied to NAICS 334416
5.50% of SIC 3661 is applied to NAICS 334418
46.31% of SIC 3679 is applied to NAICS 334418
40.75% of SIC 3679 is applied to NAICS 334410
74.19% of SIC 7379 is applied to NAICS 541512
21.22% of SIC 7379 is applied to NAICS 541519

---

[18] These coefficients apply to employment and thus differ from those shown above for the Annual Survey of Manufactures data which applied to shipments. Source: US Census Bureau, "Bridge Between NAICS and SIC", EC97X-CS3 (June 2000),

2.8.3   Updating the County Business Patterns Data

The most recent CBP data currently available is 2000. However, the Bureau of Labor Statistics also carries out a monthly "current employment survey" referred to as the "CES series". We have used the CES series to extend the CBP data to 2001 and 2002. In cases where the CES data were more aggregated than the CBP data, we applied the growth rate of the CES aggregate to each of the CBP components. Further information on the CES series is available at http://www.bls.gov/ces/cestips.htm .

2.8.4   Goods Trade Data, US International Trade Commission (ITC) (Tables 2-9 and 2-10; Figures 2-6 and 2-7)

The International Trade Commission (ITC) is the US government agency that administers a number of tasks related to US tariff rates. In conjunction with these activities, the ITC maintains a highly accessible and current database of US trade flows by 4-digit NAICS codes (since 1997). The year 2000 US data in Tables 2-9 and 2-10 come directly from this source; see http://dataweb.usitc.gov/.

The time series data in Figures 2-6 and 2-7, however, raise one further issue. ITC data from 1997 forward are provided in NAICS coding. Data for 1997 and prior years were provided in SIC coding. We have used the data for 1997 to compute a conversion ratio from SIC to NAICS for the years prior to 1997. Separate conversion ratios were computed for exports and imports and for NAICS 3341 and 3344 as follows:

(NAICS 3341 imports) = (.970)(SIC 357 imports)
(NAICS 3341 exports) = (.781)(SIC 357 exports)
(NAICS 3344 imports) = (1.027)(SIC 367 imports)
(NAICS 3344 exports) = (1.287)(SIC 367 exports)

2.8.5   Services Trade Data, US Bureau of Economic Analysis (BEA) (Figure 2-8)

The Bureau of Economic Analysis (BEA), within the US Department of Commerce, is the primary source of data for US trade in services. The data are released though annual articles published in the *Survey of Current Business* (SCB). The most recent data were obtained from:
"US International Services" (hereafter IS) in October 2002 SCB;
"US Direct Investment Abroad" (hereafter DIA) in September 2002 SCB.
For historical data, see citations in the above references, or www.bea.gov.

Figure 2-8 relies on three categories of data on trade in services: high-tech royalties, high-tech foreign direct investment earnings, and net sales of high-tech services. The following describe each of these data series.

2.8.5.1 High-tech royalties

Starting in 1997, IS has annually published royalties earned and paid on "general use computer software" with unaffiliated foreign entities (hereafter "unaffiliated computer royalties"). Historical data are also available on the aggregate royalties with unaffiliated foreign entities. We computed "unaffiliated computer royalties" for year prior to 1997 by applying the 1997 ratio for (unaffiliated computer royalties)/(unaffiliated aggregate royalties) to unaffiliated aggregate royalties for years prior to 1997. Data are also available on the aggregate royalties for *affiliated* foreign entities. To compute "affiliated computer royalties", we applied the 1997 ratio (unaffiliated computer royalties)/(unaffiliated aggregate royalties) to affiliated aggregate royalties for years prior to 1997. Total high-tech royalties is then the sum of affiliated and unaffiliated computer royalties. These computations were carried out separately for high-tech royalties received and high-tech royalties paid, and the "high-tech net royalties" equals high-tech royalties received minus high-tech royalties paid.

2.8.5.2 High-tech foreign direct investment earnings

DIA provides annual data on foreign direct investment income earned on (a) computer and office equipment and (b) computer and data processing services, representing "hardware" and "software" earnings respectively. The sum of the two series equals "high-tech foreign direct investment earnings".

2.8.5.3 Net sales of high-tech services

IS provides historical data on net computer and data processing service sales to *unaffiliated* foreign entities (hereafter unaffiliated computer sales) as well as unaffiliated aggregate sales. IS also provides data since 1997 on *affiliated* computer sales, as well as aggregate sales to affiliated foreign entities (hereafter aggregate affiliated sales). We have computed data prior to 1997 for affiliated computer sales by applying the 1997 ratio, (unaffiliated computer sales)/(unaffiliated aggregate sales), to aggregate affiliated sales for years prior to 1997. Net sales of high-tech services are then the sum of net computer sales to affiliates and unaffiliated entities.

Chapter 3

International Trade and California Industry
A Case Study of the Computer Cluster

Chapter 2 has used US and state level data to describe the aggregate interaction of global factors with a high-tech economy. In this chapter, we use industry case studies to examine production structure, the employment base, and trade flows at the firm level, thus expanding the understanding of global linkages provided by the aggregate data. Our analysis specifically looks at the computer cluster in California. In this industry, firms operating in California rely on a global resource base and production structure to produce revenues that come from markets worldwide.

3.1 GLOBAL INTEGRATION FROM A FIRM'S POINT OF VIEW: A TYPOLOGY

A firm may participate in a global economy in several ways, depending on firm size (defined by employment, revenues, and/or the number of branches), production process, market location, production and marketing networks, and, to some extent, the age or maturity of the firm[1] and personal connections and preferences of the owners and managers. The type and degree of participation may range from a simple, export oriented approach (Stage I) to a complex, multi-national structure that includes multi-directional flows of goods and services, with employment and plant ownership spread globally (Stage III).

---

[1] The variation in location decisions over the course of the firm's lifetime is the subject of the product cycle theory. See, for example, the summary in Shove (1996), and more detailed descriptions in Markusen (1985), Malecki (1979) and (1985), and Miller (1989).

3.1.1   Stage I—Export Base

Traditional regional economics assumes a local firm is a distinct entity, located within the region, employing local labor (or labor that migrates to work in the region). Sales may be within or outside the region, and a portion of sales may be outside of the country.[2] (See Figure 3-1.) Inputs may be purchased locally or imported, either from other parts of the country or from overseas. Overseas markets may affect firms either through the demand or supply side. On the demand side, changing income and tastes overseas will affect the local firm's ability to export its product. On the demand or supply side, overseas production (by foreign firms or multinationals) may have a negative or positive effect on the local firm. Foreign production of a competing product may cut into sales both abroad and in the local market (import competition). Purchases of foreign inputs may reduce costs and improve productivity for the local firm.

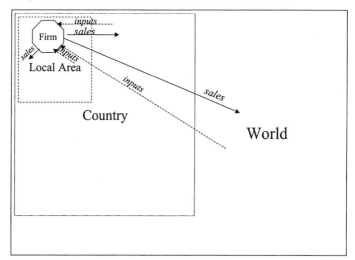

Figure 3-1. Stage I—Export Base

3.1.2   Stage II—Supply-Side-Driven Foreign Production

A firm may first transfer some or all of its own production to foreign locations to reduce production costs while maintaining control over the supply of inputs and the production process. (See Figure 3-2.) The motivation for seeking low-cost overseas production and the type of low-cost inputs or pro-

---

[2] This model of the regional economy is described in many basic textbooks. See, for example, Blair (1991, Pages 151-173).

duction sought will vary with the type of industry and characteristics of the particular firm. Factors motivating participation in the global economy include access to raw materials, low-cost labor, and overseas production network clusters.

Stage IIA—Raw Materials Variant—Firms for which raw materials are a significant part of the product (agricultural firms, oil, gas and minerals), for reasons of cost, availability, and environmental regulation, may begin to locate a portion of production overseas. This may occur either to continue to meet demand and compete in domestic markets (a response to high costs of materials or shortages domestically), or to reduce costs in existing foreign market areas, especially where perishable products make travel time a consideration.

Stage IIB—Labor Cost Variant—Firms for which labor costs in production are significant may seek a low-cost production location outside of the US. This may also occur to satisfy requirements for skilled labor. The goal of overseas production may be to provide products for sale to either foreign or domestic markets. Import competition is likely to be a spur to overseas production, as domestic firms try to lower costs in the face of competition from firms producing in a lower labor-cost environment. Intense domestic competition may also spur the effort to lower production costs through moving routine production to overseas sites.

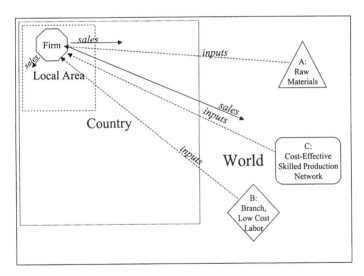

Figure 3-2. Stage II—Supply-Side Driven Foreign Production

Stage IIC—Production Network Variant—Cost-related factors may combine with other production-related concerns to determine where firms locate overseas production sites. Sites that combine sophisticated production techniques with access to suppliers and to lower cost labor become important location sites for technology-dependent firms. Singapore and Taiwan are examples of locations that have become centers for high-tech hardware manufacturing linked to US firms (and to foreign producers, as well). India and Russia are becoming centers of software production and technical expertise.

3.1.3   Stage III—Production to Market

As firms begin to sell large amounts in overseas markets, they may find advantages beyond lower costs to producing close to those markets. Under this model, establishment of foreign production sites would be determined by the location of markets as well as by the location of raw materials or low-cost labor. Products may be sold from these production sites directly to the foreign market, without passing through the domestic production or distribution facilities at all. (See Figure 3-3) The function of the domestic headquarters may then become more and more specialized in administration and research and development. Foreign sales of products from overseas plants generate income and employment at the headquarters through the repatriation of profits and the need for administrative and professional labor.

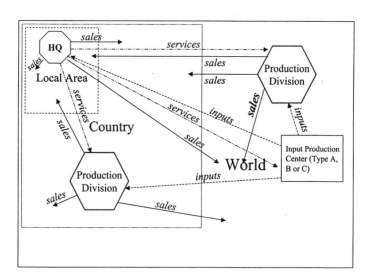

Figure 3-3. Stage III—Production to Market

As firms enter *Stage II* or *Stage III* in their production process, measures of exports and imports will no longer be sufficient to reflect the effect of foreign markets on domestic production and employment levels. A rise in sales outside the US from a foreign affiliate or branch of a California multinational firm will not show up in US export numbers, but these sales may still create a rise in the firm's domestic employment and output, even though the transactions never cross US soil.

Besides the three stages of globalization described here, global markets may affect a firm's ownership and sales opportunities in other ways. An ownership change may suddenly bring a locally oriented firm into the global market, if the new owner is a foreign corporation seeking the resources provided by the US firm. An increasingly global focus of a major customer also may quickly bring a firm reliant on local sales into the global marketplace.

There is a wide degree of variation within an industrial sector in the stage of globalization of firms. There is also wide variation among industrial sectors in the degree to which firms within each sector tend to be linked to global markets.

Case studies of different high-tech sectors illustrate the degree to which individual firms participate in the trend towards global production. Information on firm structure, location decisions, and the flows of firm inputs and outputs across borders, drawn from annual reports, web sites, and interviews of firm representatives, shows the direct and indirect paths by which industry globalization may influence the state's economic development.

3.2 THE CALIFORNIA COMPUTER CLUSTER IN OVERVIEW

California accounts for more than 20% of US employment in computer and electronic equipment manufacturing and for more than one-fourth of the nation's employment in software publishing. As discussed in Chapter 2, for much of the past decade, there has been a mismatch between growth in sales and growth in employment in the hardware segment of the cluster. Here, we expand on the most recent period, drawing from data reported for NAICS codes.

Overall manufacturing employment increased slightly in California in the 1997-2000 period, while computer-related manufacturing experienced a decline. Yet measured by shipments, the computer industry was expanding more strongly than many other manufacturing sectors (or manufacturing as a whole) during this period. Specifically, computer and electronic components shipments from California grew by 28.5% from 1997-2000, compared to the state's overall growth in manufacturing shipments of 18%. California also outpaced the nation in the growth of computer and electronic components shipments, as shown in Chapter 2 and Figure 3-4.

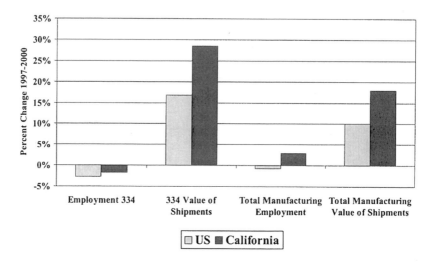

Figure 3-4. Growth in Employment and Shipment Value, 1997-2000, US and California Computer and Electronic Component Manufacturing (NAICS 334). Source: US Bureau of the Census (2000) and US Bureau of the Census (2002a).

The computer cluster also experienced strong productivity gains during this period, especially in California. In earlier research, we noted more than a doubling of value added per employee in SIC 357 between 1987 and 1995.[3] Using the NAICS categories, for the more recent 1997 to 2000 period, it is evident that further increases have occurred. For NAICS 334, value added per employee grew from $164,255 in 1997 to $217,891 in 2000, an increase of 32.7%.[4] Nationally, productivity and its growth in the computer and electronic components sector has been lower, with value added per employee reaching $177,043 in 2000, an increase of 18.5% from $149,384 in 1997. Similar data are not available for the services portion of the computer cluster. Data from the 1997 Economic Census shows that revenues per em-

---

[3] As noted in Chapter 2, SIC 357 includes much of what is in NAICS 3341. The earlier results are reported in Kroll and Kirschenbaum (1998).

[4] These numbers are not corrected for inflation, partly because of the complexities of doing a correction with an aggregate industrial category. The CPI would not appropriately reflect changes in production, as the CPI rose during these periods, but the producer price index was dropping for most sectors included in the 334 category. (See Chapter 2.) For example, electronic computer producer prices dropped by 58.9% between 1997 and 2000, personal computer producer prices dropped by 68.6%, and electronic component producer prices dropped by only 6.6%, while the CPI grew by 7.3% during the same period.

ployee in California computer cluster services tend to be higher than the national levels, as shown in Figure 3-5. Using SIC codes, growth in receipts per employee can be compared with the 1992 period. From 1992 to 1997, receipts per employee grew much faster for the computer-related services sectors (tracked through SIC codes), than for services overall (Figure 3-6). In California, receipts per employee in the computer programming and data processing sector (SIC 737) grew twice as fast as the overall growth in receipts per employee for the services sector.

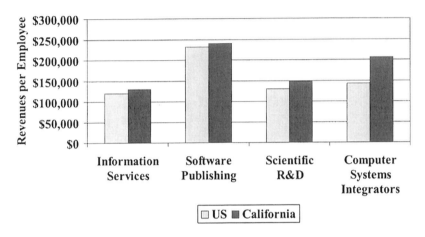

Figure 3-5. Revenues per Employee, Selected Computer-Related Services Sectors. Data is for 1997. Source: US Bureau of the Census (2000).

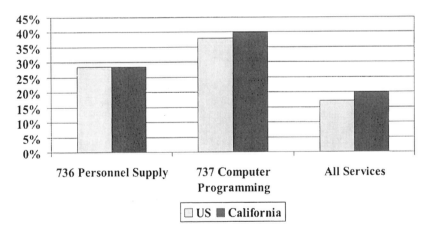

Figure 3-6. Growth in Receipts per Employee, 1992-1997, California. Source: Computed from data in US Bureau of the Census (2000), services web pages.

Globalization and a High-Tech Economy

Table 3-1. Average Hourly Wage, Manufacturing Sectors and Selected Services Sectors, 1999

NAICS/ SIC	Code Description	Wage, CA	Indexed to CA Mfg.	Wage, US	Indexed to US Mfg.
	All Manufacturing	$ 14.14	1.00	$ 14.70	1.00
311	Food mfg	$ 12.10	0.86	$ 11.75	0.80
312	Beverage & tobacco product mfg	$ 18.57	1.31	$ 18.98	1.29
313	Textile mills	$ 9.91	0.70	$ 11.09	0.75
314	Textile product mills	$ 8.95	0.63	$ 9.86	0.67
315	Apparel mfg*	$ 8.08	0.57	$ 8.60	0.59
316	Leather & allied product mfg	$ 8.64	0.61	$ 9.77	0.66
321	Wood product mfg	$ 11.43	0.81	$ 11.82	0.80
322	Paper mfg	$ 15.64	1.11	$ 16.96	1.15
323	Printing & related support act.	$ 14.67	1.04	$ 14.54	0.99
324	Petroleum & coal products mfg*	$ 25.99	1.84	$ 22.17	1.51
325	Chemical mfg	$ 16.16	1.14	$ 19.20	1.31
326	Plastics & rubber products mfg	$ 11.77	0.83	$ 12.79	0.87
327	Nonmetallic mineral product mfg	$ 14.83	1.05	$ 14.58	0.99
331	Primary metal mfg	$ 14.09	1.00	$ 17.57	1.20
332	Fabricated metal product mfg	$ 14.17	1.00	$ 14.03	0.95
333	Machinery mfg	$ 15.17	1.07	$ 15.77	1.07
334	**Computer & elec. prod. mfg***	**$ 18.08**	**1.28**	**$ 16.02**	**1.09**
3341	**Computer & periph. eq. mfg***	**$ 22.07**	**1.56**	**$ 17.05**	**1.16**
3342	**Communications eq. mfg***	**$ 20.72**	**1.47**	**$ 18.11**	**1.23**
3343	**Audio & video equipment mfg***	**$ 12.39**	**0.88**	**$ 12.73**	**0.87**
3344	**Semiconductor & oth. elec. comp. mfg***	**$ 16.18**	**1.14**	**$ 14.45**	**0.98**
3345	**Instruments mfg***	**$ 20.95**	**1.48**	**$ 18.13**	**1.23**
3346	**Magnetic & optical media***	**$ 14.56**	**1.03**	**$ 13.89**	**0.95**
335	Electrical eq., appl., & comp. mfg	$ 12.35	0.87	$ 13.75	0.94
336	Transportation equipment mfg	$ 18.63	1.32	$ 20.10	1.37
337	Furniture & related product mfg*	$ 10.87	0.77	$ 11.45	0.78
339	Miscellaneous mfg*	$ 15.07	1.07	$ 12.40	0.84
	Computer-Related Services				
737	**Computer programming, data proc. and oth. comp. rel. serv.***			**$ 22.32**	**1.52**
7371	**Computer programming serv.***			**$ 25.46**	**1.73**
7373	**Computer integrated design***			**$ 21.44**	**1.46**
7375	**Information retrieval***			**$ 15.71**	**1.07**
7378	**Computer maint. and repair***			**$ 17.29**	**1.18**

Note: All 334 sectors and high-tech services sectors in bold. * indicates sectors with California location quotients greater than 1 (737 sector LQ's are from 1997).
Source: Manufacturing wages--Annual Survey of Manufacturers; Services wages--Bureau of Labor Statistics web site.

High ratios of value added or receipts per employee and the growth of these ratios over the past decade have helped maintain high wages in the computer cluster. Production workers in computers and peripherals manufacturing (NAICS 3441), communications equipment manufacturing (NAICS 3342), and instruments (NAICS 3345) earn about 50% above the average manufacturing wage in California. The differential compared to other manufacturing workers is greater in California than in other parts of the US. Although the average California manufacturing worker earns almost 4% less than the average US manufacturing worker (based on calculations from the 1999 Annual Survey of Manufacturers, shown in Table 3-1), the hourly average wage in NAICS 334 is more than 10% above wages in that sector in the US as a whole, and computer and peripheral equipment production worker wages (NAICS 3341) in California exceed the US level by almost 30%.

Computer cluster employment tends to concentrate in specialized locations and to some extent in large firms. A significant portion of the concentration is in California. In contrast, total manufacturing employment is spread quite evenly across the states in the US, with no state having a manufacturing location quotient above 1.67 and only eight having location quotients below 0.5. (Table 3-2 shows location quotients by state for manufacturing, NAICS 334 and NAICS 3341). For NAICS 334 (all computer and electronic manufacturing), six states have location quotients above 2.0, and an additional three states (including California) have location quotients below 2.0 but above 1.67, the highest location quotient for manufacturing overall. Location quotients are below 0.5 in nineteen states. While California does not have the highest location quotient, it has the largest concentration, by absolute number, of any state, with over 20% of US employment in NAICS 334. About half of US employment in 334 is in six states—California, Texas, Massachusetts, New York, Illinois and Minnesota. The degree of agglomeration is even higher in the more specialized 4-digit sectors. Nine states have location quotients above 2.0 in NAICS 3341, and thirty-one states have location quotients less than 0.5 (including some large states with strong manufacturing bases). Almost one-fourth of US employment in NAICS 3341 is concentrated in California and close to 50% of the sector's US employment is in just four states—California, North Carolina, Texas and Minnesota.

Table 3-2. State Location Quotients in Total Manufacturing, NAICS 334 and NAICS 3341

Area Name	Manufacturing Emp.	334 Emp.	3341 Emp.	LQ-Manufacturing	LQ-334	LQ-3341
US	16,473,994	1,557,087	193,897	1.00	1.00	1.00
Alabama	333,782	15,109	4,500	1.40	0.67	1.60
Alaska*	11,258	60	10	0.38	0.02	0.03
Arizona	200,859	53,116	1,544	0.72	2.03	0.47
Arkansas*	235,578	4,687	10	1.65	0.35	0.01
California	1,753,716	335,662	45,211	0.94	1.91	2.06
Colorado	166,461	37,603	7,672	0.60	1.44	2.36
Connecticut	232,789	21,494	730	1.04	1.02	0.28
Delaware*	41,846	2,152	60	0.77	0.42	0.09
District of Columbia*	2,613	60	-	0.04	0.01	0.00
Florida	415,435	57,731	2,807	0.46	0.68	0.27
Georgia	518,063	16,225	2,407	1.03	0.34	0.41
Hawaii*	14,844	375	-	0.24	0.06	0.00
Idaho	67,103	17,110	3,242	1.03	2.78	4.23
Illinois	852,646	69,440	2,539	1.07	0.92	0.27
Indiana*	639,185	26,427	375	1.67	0.73	0.08
Iowa*	244,806	11,527	375	1.34	0.67	0.17
Kansas	191,609	6,062	848	1.18	0.39	0.44
Kentucky*	293,736	9,759	7,500	1.34	0.47	2.91
Louisiana*	161,415	2,320	10	0.70	0.11	0.00
Maine*	79,645	5,681	10	1.12	0.85	0.01
Maryland*	158,753	24,357	750	0.53	0.87	0.21
Massachusetts	397,620	102,854	6,157	0.89	2.44	1.17
Michigan	819,227	24,123	1,185	1.39	0.43	0.17
Minnesota	377,690	58,758	15,439	1.09	1.80	3.79
Mississippi*	220,046	4,560	375	1.59	0.35	0.23
Missouri	347,772	12,051	791	1.00	0.37	0.19
Montana*	21,019	527	10	0.49	0.13	0.02

Area Name	Manufacturing Emp.	334 Emp.	3341 Emp.	LQ-Manufacturing	LQ-334	LQ-3341
Nebraska*	108,642	6,323	10	1.00	0.62	0.01
Nevada	38,146	3,247	254	0.29	0.26	0.17
New Hampshire	93,048	18,547	1,293	1.18	2.49	1.39
New Jersey	386,095	34,393	1,632	0.75	0.71	0.27
New Mexico*	38,072	9,944	175	0.48	1.33	0.19
New York	705,914	74,354	9,883	0.66	0.74	0.79
North Carolina	731,399	45,731	17,230	1.50	0.99	2.99
North Dakota*	24,025	1,672	10	0.65	0.48	0.02
Ohio	988,612	33,721	2,471	1.37	0.49	0.29
Oklahoma	168,580	11,316	2,191	0.97	0.69	1.07
Oregon	202,657	33,712	9,876	1.04	1.82	4.29
Pennsylvania	798,333	54,942	2,415	1.09	0.79	0.28
Rhode Island	68,558	7,142	788	1.14	1.26	1.12
South Carolina*	334,651	14,537	375	1.45	0.66	0.14
South Dakota*	46,675	10,924	7,500	1.05	2.61	14.38
Tennessee	475,621	14,906	1,849	1.38	0.46	0.46
Texas	966,396	133,267	17,127	0.83	1.22	1.26
Utah	122,542	15,752	3,129	0.93	1.26	2.01
Vermont	45,217	10,294	-	1.23	2.97	0.00
Virginia	360,237	31,594	4,542	0.86	0.80	0.92
Washington	315,102	45,554	5,123	0.96	1.47	1.33
West Virginia*	74,209	912	175	0.92	0.12	0.18
Wisconsin	572,060	24,210	3,413	1.64	0.73	0.83
Wyoming*	9,687	413	175	0.38	0.17	0.59

*Employment estimated by authors from other County Business Patterns data. Note: Location quotients differ slightly from those in Chapter 2 because they are based on a different data source. Source: Authors from US Bureau of the Census, *County Business Patterns 2000* data.

The manufacturing portion of the computer cluster is concentrated in larger firms as well as by location. Statewide, 14% of manufacturing employment is concentrated in establishments with 1,000 or more employees. In NAICS 334, 31% of California employees are in firms of this size. At the US level, the industry is even more concentrated in large firms, with 38% of US NAICS 334 employment in firms of 1,000 or more employees. (This is consistent with the location of administrative and R&D functions near the headquarters of California firms and the spread of more routine, larger manufacturing facilities to lower cost areas out-of-state.)

The California software and services portion of the cluster are less heavily concentrated in large firms, but are more concentrated than are US software and services firms overall. The California information sector (NAICS 51) has 17% of employees in large firms of over 1,000 employees; 18% of information and data processing employees (NAICS 514) and 20% of employees in professional, scientific and technical services are in firms in this size category (NAICS 541). In contrast, 15% of US employees in the information sector and the information and data processing sector, and 10% of professional, scientific and technical services employees are in firms of 1,000 or more employees.[5]

As noted in Chapter 2, the computer cluster has high import and export flows. In 2000, more than half of the total domestic sales in the computer and peripheral manufacturing sector (3341) were of imported goods, showing both the reliance on imported inputs and the strong level of import competition the industry faces. At the same time, 40% of the value of shipments from US computer and peripheral manufacturing firms was exported, indicating the importance of international markets to these firms.

Within California manufacturing sectors, there is a wide range in the importance of trade (either as exports, imports, or both), in the direction of trade balance, and in the direction and share of employment change. Tables 3-3 and 3-4 show the computer cluster's broad pattern of high trade flows, unbalanced trade, weak employment growth and strong shipment growth. Even within the computer cluster, sectors sharing similar trade patterns have differing levels of employment growth, although both 3341 and 3344, as well as closely related 3342, had shipment growth far stronger than most other sectors in the state. [6]

Facing both competition from foreign producers and opportunities in foreign markets, the computer cluster in California is well positioned for a global role. International sales (either exports or sales outside the US by

---

[5] Calculated by authors from *County Business Patterns 2000* data.

[6] The case study included networking firms, which may encompass activities not only in 3341 and 3344 but also in 3342.

overseas branches) may grow even as domestic sales are reduced by competition from foreign producers in US markets, while the high cost of production in California makes the search for alternative production sites attractive. At the same time, the advantages of agglomeration and firm concentration keep a portion of the industry headquartered in California.

Table 3-3. Comparison of California Shipment Growth 1997-2000 with US Trade Balance (Manufacturing Sectors with 25,000+ Employees; Percent Change in Parentheses)

	Shipment Growth Negative	Shipment Growth Positive
Positive Trade Balance	3364 Aerospace Pr/Parts (-25%) 3231 Printing (-2%)	3345 Instruments (10.4%) 3261 Plastics Products (14.6%) 3339 Gen. Purpose Machinery (4.2%) 3391 Medical Eq. (43.7%) 3222 Paper Products (17.3%)
Negative Trade Balance		**3344 Semiconductors (78.3%)** 3152 Cut and Sew Apparel (11.4%) **3342 Communications Eq. (77.2%)** **3341 Computer Eq. (73.9%)** 3327 Machine Shops (16.1%) 3399 Misc Mfg. 22.7%) 3371 Furniture (18.6%) 3323 Arch/Str Metals (20.4%) 3118 Bakeries (21.5%) 3114 Fruit and Vegetables (4.9%) 3254 Pharmaceuticals (50.3%) 3219 Other Wood Products (13.4%) 3332 Industrial Machinery (78.5%) 3363 Motor Vehicle Parts (15.1%) 3329 Other Fabricated Metal (2.2%)

Note: Shipment growth adjusted for inflation using PPI by sector. **Computer cluster and related sectors in bold.**

Source: Shares computed by authors from *Annual Survey of Manufactures* and international trade statistics described in Chapter 2.

Table 3-4. Comparison of California Employment Growth 1997-2000 with US Trade Balance

	Employment Growth Negative	Employment Growth Flat or Positive
Positive Trade Balance	3364 Aerospace Products/Parts*√ (-28.8%) 3345 Instruments*√(-1.9%)	3261 Plastics (14.5%) 3231 Printing. (2.7%) 3391 Medical Eq.*√(25.2%) 3222 Converted paper pr. (1.8%) 3339 Other gen. pur. mach.√(0.0%)
Negative Trade Balance	3152 Cut & Sew Apparel*√(-5.0%) **3341 Comp. & Per. Eq.*√(-35.7%)** 3114 Fruit & Veg Pres.* (-8.4%) 3363 Motor Veh. Parts √(-2.1%)	**3344 Semiconductors*√(11.3%)** **3342 Commun. Eq.*√(9.9%)** 3327 Machine shops (13.4%) 3399 Other Misc. Mfg.√(20.2%) 3371 Furniture (22.0%) 3323 Arch. & Str. Metals (5.6%) 3118 Bakeries & tort. mfg. (10.8%) 3254 Pharmaceuticals (29.3%) 3219 Other Wood Pr. (10.7%) 3332 Industrial Machinery√(22.1%) 3329 Other Fab. Metals√(2.7%)

* Location Quotient greater than 1.5. **Computer cluster and related sectors in bold.**
√High trade-flow sector.
Source: Shares computed from *Annual Survey of Manufactures* and international trade statistics described in Chapter 2. Location quotients based on County Business Patterns data.

3.3   RESEARCH APPROACH AT THE FIRM LEVEL

To understand the forces behind globalization of the industry and the pattern of that globalization in more detail, the authors conducted research at the firm level on several segments of the computer cluster. The research was initiated with interviews in 1997 and was updated in 2002. Firms were identified both initially and during the update primarily from the *Hoover's Guide to Computer Companies* (1996) and *Hoover's On-Line.*

We identified firms for the initial interviews and for the later Internet research based on the following categories:[7]

---

[7] The sample we drew from Hoover's uses different groupings into sub-sectors of the computer cluster than described in Chapter 2. The firms included in our definitions all fall within NAICS 334, 51, or 54, but an individual grouping does not draw exclusively from one sector, and groupings may include additional sectors related to those discussed in Chapter 2.

Diversified Computer Hardware: These are the firms that actually produce integrated computer systems (Hewlett Packard, Sun Microsystems, Apple Computer, for example). These are companies primarily identified by NAICS 334111, Electronic Computer Manufacturing.

Computer Components and Peripherals: These are computer cluster manufacturers who make computer components such as storage devices or peripheral equipment such as monitors. They are identified by several NAICS codes, primarily 334112 (Computer Storage Device Manufacturing), some firms within 334418 (Printed Circuit Assembly), and 334419 (Other Electronic Component Manufacturing).

Semiconductors: These firms manufacture semiconductors and integrated circuits (NAICS 334412 and NAICS 334413 primarily). Some provide design services as well (NAICS 5415).

Networking and Communications Devices: These firms provide both the hardware (including components) and software critical to communications between computer systems. This includes firms from NAICS 334119 (Other Computer Peripheral Equipment Manufacturing), 33421 (Telephone Apparatus Manufacturing), 334413 (Semiconductor and Related Device Manufacturing), 334418 (Printed Circuit Assembly), and 5415 (Computer Systems Design and Related Services). Distributors focusing entirely on networks are also included (NAICS 42143).

Software Firms: These firms design computer software. Many sell packaged products but some design custom products or provide consulting on the selection and use of their products. Primary industrial codes include NAICS 51121 (Software Publishers), 541511 (Custom Computer Programming Services), and 541512 (Computer Systems Design Services).

Services Firms: With the growth of computer hardware and software has come the industry that assists users of the products. This grouping includes firms that provide system consulting, specialized software engineering, integrated software applications systems, and other information technology services. The primary industrial code is NAICS 5415 (Computer System Design and Related Services). The grouping also includes firms in NAICS 5141 (Information Services), 5142 (Data Processing Services), and 5112 (Software Publishers).

We conducted detailed interviews in 1997 of firms in four of these groupings, Diversified Computer Hardware companies, Computer Components and Peripherals manufacturers, Networking and Communications Devices companies, and Software firms. In 2002, we updated information on most of the firms interviewed, primarily using material obtained from web sites, annual reports, and articles in the business press. (Some firms had ceased doing business in California or had merged with other firms). As shown in Table 3-5, the interviews covered fourteen firms in depth and six other firms in less detail. Some 90 additional firms are covered through re-search using published materials. Firms selected for interviews were either the largest or fastest growing firms in the sector. The additional 90 firms added to the sample represent firms headquartered in California with 500 or more employees world-wide, and include services and semiconductor firms as well as firms in the initial four categories.

From this broad sample of firms, we were able to examine variations in characteristics related to global production and sales. Key factors examined include:

- Firm size
- Age of firm
- Geographic distribution of production
- Geographic distribution of markets (sales)

These characteristics varied among industry groupings and within group-ings. Characteristics of firms, and of a general grouping, also changed over time, as industries evolved from their earliest start-up years and as the ease of communications increased.

Table 3-5. Research Coverage of Firms in Computer Cluster Groupings

Computer Cluster Grouping	Initial Interview Subjects (1997)			Web Research (2002)	
	Firms in Sample	Completed Interview	Annual Report Data or Partial Response	2002 Update	Additional Firms, 2002
Diversified Hard-ware	10	5	1	3	3
Components and Peripherals	7	2	1	2	16
Networking	5	4	1	3	9
Software	13	3	3	5	27
Semiconductors	-	-	-	-	27
Services	-	-	-	-	8
Total	**35**	**14**	**6**	**13**	**90**
Source: Authors' research 1997 and 2002.					

3.4   VARIATIONS IN BASIC STRUCTURAL CHARACTERISTICS

Some very large computer hardware firms have been leaders in establishing a global presence. Several of these firms employ tens of thousands of workers worldwide. Examples in the diversified grouping are Hewlett Packard, with world-wide employment of 86,000 in 2001 and Sun Microsystems, with 43,700. Although somewhat smaller, Apple Computer, with 9,600 employees, Gateway with 14,000, and Silicon Graphics with 6,000, further emphasize the role of large players in the production of computers. There are few small to mid-sized computer hardware manufacturers in California (as measured by world-wide employment).

In the peripherals and components grouping, Seagate Technology employed 60,000 world-wide in 2001. Eight other firms producing peripherals and components employed between 2,000 and 8,000 in 2001. In Networking, Cisco systems employed 38,000 worldwide in 2001. 3Com, with 8,000, and Adaptec and Xircom, with close to 2,000 each, are the only other large firms in this category. This grouping has a much larger share of small to mid-sized firms, in the range of 200 to 1,000 employees. Nevertheless, the larger firms dominate employment and revenues overall.

Five California-headquartered semiconductor companies employed 10,000 or more in 2001, including Intel (with 83,000), Solectron (with 60,000), Applied Materials, Advanced Micro Devices, and National Semiconductor (each with 10,000-20,000 employees). Another seventeen firms have between 1,000 and 10,000 employees worldwide.

The software and services sectors support fewer giant firms, although in some cases large firms that are primarily hardware producers, such as Hewlett Packard, have significant software and service components as well. The largest software firm headquartered in California, Oracle, had 41,000 employees worldwide in 2001. An additional twenty-two firms had between 1,000 and 10,000 employees worldwide. Computer Science Corporation, the largest services firm (not including Hewlett Packard), had 66,000 employees in 2001 but very few other services firms with California headquarters had employment greater than 1,000 worldwide.

The majority of firms in our sample are publicly traded. Many used initial stock offerings as a way for raising capital in the early stages of expansion. The age range of firms varies significantly by grouping, as shown in Figure 3-7. In general, this is a young industry, with very few firms established before 1970. At one extreme, half of semiconductor firms (with 500 or more employees) were established no later than 1980. At the other extreme, more than half of both networking and services firms were established after 1990.

Globalization and a High-Tech Economy

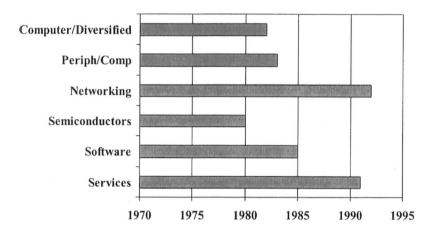

Figure 3-7. Median Startup Year of Computer Cluster Firms, by Grouping. Source: Authors, using data from Hoover's On-Line and Lexis-Nexis.

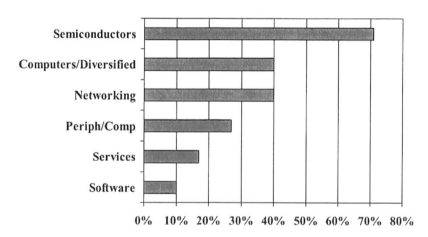

Figure 3-8. Share of Firms With More Than Half of Sales Outside US, by Computer Cluster Grouping, (Firms with 500+ Employees). Source: Computed by authors from data compiled from financial databases, web sites, and annual reports.

3.5   FOREIGN SALES AND EXPORTS

For the great majority of firms in this cluster, sales generated outside of the US are a significant proportion of revenues. While a few firms report less than 10% of revenues from foreign sales, many of the largest firms report between one-third and two-thirds of sales abroad. These sales are not necessarily produced in California or even in the US, and may not be counted as exports in the national accounts. It is common for the largest firms to do much of their production outside the US, as described in a later section of this chapter.

The share of sales occurring outside the US and the location of those sales varies significantly among cluster groupings and among firms within each grouping. In general, the manufacturing firms were far more reliant on foreign sales than the nonmanufacturing firms, as illustrated in Figure 3-8. On average, semiconductor firms are most dependent on foreign sales. Of the twenty-four firms with 500 or more employees for which foreign sales data was available, over 70% received more than half of their revenue or sales from outside the US. About two-thirds of semiconductor firm foreign sales were in Asia. This reflects the fact that the semiconductor industry is an input supplier to computer systems, and many computer systems are assembled in Asia.

Both diversified computer hardware firms and network and communications firms are also heavily reliant on sales outside the US. In both groupings, 40% of firms rely on foreign sales for more than 50% of sales or revenues. On average, over 40% of sales from larger firms in these sectors (500+ employees) occur outside the US. In both groupings, Europe accounts for a larger share of overseas sales than does Asia, as shown in Figure 3-9.

The components and peripheral manufacturing industry on average has fewer firms that sell the majority of their output outside the US, but a portion of those firms, primarily makers of components installed directly in computers, are highly dependent on foreign sales, largely to Asia. Read-Rite Corporation (headquartered in Fremont, California), for example, sold 86% of its output to five Asian countries in 2001. Komag, Incorporated (headquarters San Jose, California) relied on Malaysia and Singapore for over 90% of sales in 2000. Many of these sales are to the Asian subsidiaries of US multinational enterprises. Makers of peripherals, such as storage devices, are more likely to sell about two-thirds of their product within the United States.

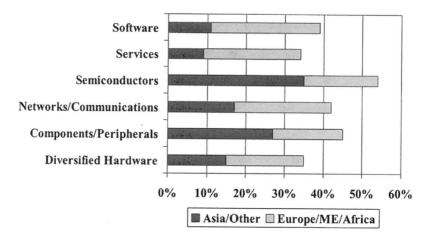

Figure 3-9. Estimates of Percent of Sales to Asia and Europe by Computer Cluster Groupings (Firms with 500+ Employees). Source: Computed by authors from data compiled from financial databases, web sites, and annual reports.

Services and software firms are less reliant on overseas sales, compared to manufacturing firms, but their reliance on overseas sales is increasing over time. In our initial research, in 1997, software firms averaged about 10% of sales overseas. The data for 2000 and 2001, instead, indicates that on average overseas sales account for more than one-third of sales of California headquartered software firms (with 500+ employees), and half of firms sell at least 40% of their product overseas. Services firms average 30% of sales overseas, and half of services firms sell at least one-third of their product in foreign markets. Foreign sales of software and services firms are primarily to European rather than Asian markets. The increasing importance of global sales in the nonmanufacturing segments of the computer cluster are indicative of a number of changes that have taken place in the past decade. The penetration of computer products in foreign markets and the globalization of information networks via the Internet and the World Wide Web have greatly increased demand for computer software and services outside the US, and the larger software and services firms see international markets as a fertile area for new growth.

3.6   FOREIGN PRODUCTION

With the growth of overseas sales has come increasing production activity outside the US, in both manufacturing and non-manufacturing segments of the computer cluster. A simple model of export activity does not apply to the majority of firms in the computer cluster. Most firms combine elements of Stage II and Stage III in their production process. In some groupings of the computer cluster, firms aggressively seek low-cost sites, and maintenance of low-cost production facilities is crucial to their continued competitiveness. In other groupings, proximity to markets is a much stronger driving factor. Some firms combine the search for low-cost production with the move to new markets.

Semiconductor firms have a long history of locating routine production in Asia.[8] Some firms, such as Applied Materials, LSI Logic, and ChipPAC became involved in production overseas in the late 1970s and the 1980s. Other firms, for example Solectron and Maxim Integrated Products, have gone through major waves of expansion of overseas sites in the 1990s. A great deal of variety in ownership patterns characterizes foreign manufacturing operations of these firms. Some are owned as subsidiaries, either constructed for that purpose or, at least as often, purchased as an already operational facility from a competitor or customer. In some cases, not just the plant but an entire company has been purchased to give the firm entrée into a new market niche as well as a production facility.

A second pattern of ownership of foreign plants is through a joint venture operation with a foreign company. A few smaller semiconductor firms have maintained their small size by contracting out manufacturing to overseas firms. Both the joint venture and contracting has occurred mainly with firms in Asia, often with those located in countries with large agglomerations of high-tech manufacturing facilities.

While the majority of overseas activity for semiconductor firms is either manufacturing or sales, more technical development activity is also taking place overseas. In some cases, this is through acquired, specialized overseas firms. In other cases it is through the establishing of research and development or design departments at sites where manufacturing is also occurring. Some firms, such as LSI Logic (in the United Kingdom) and ChipPAC (in Korea) established design centers at overseas sites as early as the 1980s. The location decisions for these firms combine reasons from both Stage II and Stage III of the globalization process. Many of the production sites have been picked for low costs and production network advantages (Stage II). The Asian location of many of these plants also reflects the fact that many customers also do their production in Asia. Recent interest in expansion of plants into China reflects a combined interest in a low-cost, skilled labor force and the possibility of rapidly expanding demand in China for products using semiconductors.[9]

---

[8] See for example Saxenian (1994).
[9] Cole (2002).

Components and peripherals manufacturers also have extensive networks of foreign manufacturing plants, especially in Asia. Malaysia, Singapore, Taiwan and Thailand are the major sites for manufacturing plants, with many clustered around existing high-tech agglomerations. A number of these firms have also had close ownership ties to foreign firms, either through joint ventures or through acquisition.

Diversified computer hardware manufacturers were among the earliest to move production overseas. Hewlett Packard, among the oldest of the computer cluster firms in California, established its first overseas marketing (Switzerland) and manufacturing (Germany) operations in 1959 and started a Japanese joint venture in 1963.[10] Underlying reasons for these location decisions are likely different from expansion decisions in the 1980s and 1990s, with market interest and available knowledge and technical skills being of greater importance than low-cost production opportunities. Hewlett-Packard manufacturing facilities today are located throughout the world, reflecting a search for both skilled employees and market opportunities. This pattern of facility location appears with other computer manufacturers as well. Ireland and the United Kingdom are as common as Asian locations for manufacturing facilities, giving the firms easier access to European markets. Development facilities are found in Asia (primarily in Indian) and Europe as well as the US.

Networking and communications include routine manufacturing, customized manufacturing, services and software firms. No single pattern of production characterizes these firms. Some of California's largest networking firms are primarily at Stage III. Proximity to customers helps provide the technical expertise needed to implement a system. Much of the hardware is produced by suppliers, rather than manufactured in-house or by international branches. These firms have faced little foreign competition, with cost pressures coming from changes in the Internet sector, rather than from low-cost foreign production. The segment of this grouping that produces communications equipment for standard products such as personal computers has a production pattern that looks more like the component and peripheral manufacturers, with manufacturing occurring in the locations that offer specialized production networks, such as Malaysia and Singapore—a classic Stage II low labor cost variant. Other large Networking firms showed less concern with input costs, indicating in interviews that tax policy and levels were more of a burden than labor costs.

Software and services firms perhaps have undergone the greatest transformation over the past half-decade. Interviews in 1997 showed many young, rapidly growing firms were closely tied to the labor force in their

---

[10] Hewlett Packard history web page.

California location. The customer base for software firms supporting hardware was often concentrated in Silicon Valley or the greater San Francisco Bay Area. A few larger, older firms had up to half of their labor force out-of-state, with many working in other US or overseas "branches" that were firms recently acquired by the parent company. In the subsequent five years, many new start-up firms have emerged (and many have closed shop), and some of the earlier players have begun to level off in size after broad expansions in their customer base.

The global reach of software firms may take several forms. The type chosen depends in part on the character of the firm's products. For firms producing packaged software for use by individual consumers (rather than businesses), the most common global expansion is through overseas marketing offices. Accompanying this marketing move may be overseas offices that customize the software for the local market (e.g. translating into foreign languages, changing units of currency or units of measure), and technical assistance offices, that field calls from users (a type of Stage III). The most likely remote production activity is the reproduction of disks and related printing, although this may be contracted out (Stage II or Stage III, depending on location). Firms producing more complex systems for business use find a greater need for technical assistance and marketing presence near foreign markets. At Oracle, for example, consulting, support and training services account for over 60% of sales. Software companies may have dozens of international offices designed to keep their product working for major customers.

Services firms have foreign office patterns similar to the software firms producing complex products for business clients. As the firm begins to sell globally, locating staff overseas to serve clients in foreign market areas becomes an advantage. More than four-fifths of California's computer services firms with 500+ employees have at least 20% of sales outside of the US, and all of these have overseas offices.

The type of foreign direct investment described here might be expected to replace home-country exports and perhaps have a negative effect on employment opportunities in the company's headquarters region. Research has shown that the process is not so simple. Clausing (2000), for example, found that multinational activity and trade are complementary activities. These effects are further explored in Chapter 6.

3.7 IMPORTED INPUTS IN THE PRODUCTION PROCESS

Imported inputs have played a significant role in helping US computer cluster manufacturers compete with foreign firms. The 1997 interviews asked respondents about the types of inputs imported. Memory chips were the most common product imported, coming directly from Asia. Other chips, circuits and components were also imported from Asia.

The share of imported inputs in the cost of production varied widely by firm, as did the way in which inputs were imported. Respondents estimated that 10% to 20% of their inputs were imported. They added that inputs purchased from other California suppliers might also have been imported. In addition, for many firms, because of their overseas operations, intra-firm shipments from overseas branches (sometimes called transshipments or inter-area transfers) were a significant portion of their inputs. A personal computer assembled in California may contain components produced by a branch of the firm in one of its foreign locations, and a California plant may provide components to the firm's overseas manufacturing site. Firm use of transshipments varies widely among sectors within the computer cluster. At one extreme, one manufacturer of components reported transshipments from foreign (company owned) production facilities equal to over 70% of the value of shipments. Most component manufacturers report foreign transshipments equal to 50% of shipments or higher. At the other extreme, none of the software firms whose annual reports we reviewed in 1997 reported foreign transshipments. Computer manufacturers also showed significant levels of within-company foreign production of inputs, while network manufacturers, like software producers, showed very low levels of foreign transshipments.

In Chapter 4, we present evidence on the growth of imported inputs in high-tech sectors. In addition, the chapter explores the role of imported inputs in reshaping the computer manufacturing industry labor force. Chapter 5 includes further analysis of transshipments. The following two sections of this chapter more broadly examine the changing labor force, in the hardware, software and services sectors, and outsourcing as a way of tapping pools of foreign labor.

3.8   LABOR FORCE CHARACTERISTICS OF MULTIPLE-LOCATION FIRMS

Results from the 1997 interviews showed that, in most cases, the profile of employees at California locations was quite different from out-of-state or overseas employees. California locations of most firms had a much larger share of professional and technical employees than overseas locations. Only a small number of firms gave detailed breakdowns of their employee profiles. Of these, most reported that at their California locations, between 30% and 40% of employees were in professional and technical (or R&D) occupations. These shares were generally much smaller outside of California, where sites were more likely to be primarily devoted to production or sales.

While overseas operations tended to have higher proportions of production workers, many California firms maintain a production presence within the state. California producers tend to focus on high-end products, but again, strategies vary widely. At one extreme are firms that produce high-cost, complex equipment and maintain the bulk of their production within Silicon Valley. Firms that can afford this strategy are those facing little competition for their product, and where customized aspects of their production are important to their customers.[11] At the other extreme are firms that do almost all of their production at low-cost sites outside of California. Most computer and software firms lie somewhere in the middle, maintaining at least one production site in California but also establishing at least one production site overseas. Within California, a few firms have moved some or all of their domestic production to lower-cost California sites, such as Sacramento.

Because of the high share of professional and technical workers and the lower share of production workers, computer firms require a highly trained labor force in California. Advanced degrees (masters or Ph.D. level) are required of virtually all professional and technical employees, and of many management and administrative employees. A minimum of a four-year degree is required of many sales employees as well. Where products are customized, a significant share of the production workforce may also require four-year or advanced degrees, while community college or other training may suffice for the production workforce for more standardized products.

Most firms recruited their workforce from universities throughout the US and beyond. However, a small proportion drew mainly from California universities, either because of the common approach this gave their staff or because this avoided relocating new employees to a high-cost setting. Many firms rely, at least in part, on foreign-born workers for both routine tasks and skilled labor, although the level varies widely and is not accurately reported. Estimates range from over 10% in some software firms to as high as 60% in one computer/peripheral production operation.

Some conditions have changed since the 1997 interviews. The Internet and high-tech boom of the late 1990s greatly raised the demand for technical workers in California. Employment in Computer and Data Processing in California grew by 160,000 between 1997 and 2000.[12] The demand for new workers led the high-tech industry to lobby for a higher quota for temporary foreign workers under the H-1B program. The quota, which was 65,000 at

---

[11] This finding is similar to that reported by Saxenian (1994).
[12] Calculated from California Employment Development Department data, available from http://www.edd.cahwnet.gov/.

the time of the interviews, was raised to 115,000 in 1998 and to 195,000 in 2001. It is likely that the tight domestic labor market and the increase in H-1B workers increased the share of foreign workers in computer cluster firms in California in 1999 and 2000, although weaker conditions in 2001 and 2002 may have leveled off and even reversed this trend.[13]

Since the mid-1990s, there have also been further developments in the use of international labor pools of skilled workers outside the US. The following section presents examples of countries where US corporations have set up technical and professional divisions and where independent firms have evolved whose primary customer base is the US computer cluster and other high-tech firms.

3.9   INTERNATIONAL SOFTWARE AND SERVICES OUTSOURCING

The past half-decade has seen the development of production centers similar to hardware input clusters in Taiwan and Singapore but specialized in software production and information technology services. India has dominated this trend, although Russia and China appear likely to be major participants in this activity in the future, and smaller amounts of outsourcing are happening in Ireland, Israel, South Africa, and Asian locations where hardware is manufactured, such as Singapore, the Philippines and Malaysia.

India has a highly developed specialization for software and services outsourcing. India's major trade organization for the IT industry, Nasscom, reports the country's output for the software industry at $10.1 billion, of which $7.8 billion is exported. The United States accounts for almost two thirds of the exports (62%), and Western Europe for almost one fourth (24%).[14] The country's software output has increased more than five-fold since 1996, averaging a growth rate of over 40% per year. Exports account for 80% of this growth.

Several factors have allowed India to take the lead in outsourcing from the United States. First, there are many personal links that have developed from US-educated Indian entrepreneurs, professionals and technicians and from Indian immigrants to the US. The Institute of International Education (IIE) estimates that India accounted for over 10% of foreign students enrolled in US universities during the 2001/2002 academic year.[15] Research by

---

[13] Khirallah (2002).

[14] Dossani (2002), Nasscom (National Association of Software and Services Companies) web site at http://www.nasscom.org.

[15] Institute of International Education (2002).

Saxenian of Indian-born professionals in Silicon Valley found that more than half have been involved in company start-ups in the US or India.[16] Second, India can provide a well-trained labor force in engineering and software programming. Three-fourths of Indian students in the US are studying at the graduate level, according to the IIE. Third, governmental organizations, trade-organizations such as Nasscom, public educational institutions, such as the Indian Institutes of Technology, private sector technical colleges and public-private partnerships have cooperated to ensure that educational infrastructure and business support and information are available to expand the supply of the labor force as demand for outsourcing grows from the US and other countries. For example, the Indian Institute of Information Technology is a public-private partnership in Bangalore designed to support the information technology industry in Bangalore through education and research.

Fourth, the IT outsourcing industry in India has built on English language capabilities and geographic location to provide US firms with 24-hour access to English-speaking personnel. Finally, India has been a lower cost environment than the United States or Europe. Overall manufacturing wages are about 1.5% of US wages. The differential appears to be less extreme for software development, but a US firm may still save as much as 90% in labor costs by outsourcing to India, according to some sources.[17]

Despite India's early and successful entrance into software outsourcing, it faces competition from many other parts of the world. Other countries like Ireland and Israel and several Asian countries have moved in a much smaller way into outsourcing through contacts established through other types of activities. Ireland, for example, became a center of high-tech hardware production and sales distribution during the 1980s and 1990s. The country was affected by the high-tech downturn in 2001, which led to some plant closures.[18] However, the two decades of high-tech growth were accompanied by the build-up of a skilled labor force in engineering, allowing Ireland to move into software production. Ireland is the one European country with outsourcing related software exports comparable to those of India.[19] In addition to the growth of a skilled labor force, Ireland has attracted US firms as a lower cost English-speaking country with membership in the European Union.

---

[16] Saxenian, Motoyama and Quan (2002).
[17] Hackbarth (2002).
[18] Bowman (2002).
[19] Mayer (2002), Norton (2002).

Table 3-6. Comparative Economic and Demographic Data, Selected Outsourcing Countries

	Labor Force (Millions, 2000)	R&D Research- ers (Thousands, 1996/1997)	R&D Research- ers per Million Inhabitants	Hourly Manu- facturing Wage (1995 Estimate)
India	406.0	142.8	151	$ 0.25
Ireland	1.8	NA	NA	$ 9.75
Israel	2.4	NA	NA	$ 8.25
Japan	66.7	617.4	4909	$ 23.66
Malaysia	9.9	NA	NA	$ 1.59
Mexico	39.8	NA	NA	$ 1.75
Philippines	32.0	NA	NA	$ 0.71
Russian Federation	71.3	561.6	3801	$ 0.90
USA	141.8	985.5	3697	$ 17.20
European Union*	172.8	824.9	2211	$ 21.00

Sources: Labor force: World Development Indicators web; Franco and Jouhette (2001). R&D researchers: Unesco Institute for Statistics (2001, page 7). Wages (China, India, Japan, Malaysia, Philippines, USA): World Bank (2002, page 45). Wages (Ireland, Israel, Mexico, Russian Federation): International Labor Organization web site. * Wage is author estimate based on data in Nobre 1999. Note: Services wage differentials may be quite different.

Even with language, cost, skills or location advantages, India and Ireland are not always the first choice location for US or European firms seeking outsourcing. In India, political instability has raised concern with some firms,[20] and there are labor force shortages in some areas of specialized technical expertise. Firms considering outsourcing from Ireland have had to weigh the advantage of location and language against the limitations of a much smaller labor force and rising wage levels, as shown in Table 3-6. Russia and China have the potential in the long term to become major participants in software and information technology services production.

In Russia, software outsourcing is the focus of a number of firms and trade organizations. Russia has been undergoing a systemic transformation over the last decade and the economic transition has involved privatization, liberalization, creation of new institutions and the emergence of entire new branches and sectors of the economy. The slow demise of the Soviet military-industrial complex, the sectoral restructuring in the Russian economy and the prolonged economic crisis has led to a serious underemployment and underutilization of its human capital and its technical manpower. While serious efforts have been made in the arena of "military conversion" they have not borne success to any significant degree, partly due to their strategy of concentrating on manufacture of consumer durables. The emergence of the software sector, beginning in the 1990s, has offered a new opportunity for the vast dormant pool of scientists, engineers and mathematicians.

---

[20] Lingblom (2002), Vijayan and Hoffman (2002).

The traditional strength of the education system in Russia in math and the sciences finds its expression in the sophisticated algorithms and complex coding which are the hallmarks of the Russian software sector vis-à-vis the Indian. Russia has almost four times the number of "R&D researchers" as India with a labor force less than one fifth the size, as shown in Table 3-6. Although wages are higher than in India, they remain well below other outsourcing locations such as Ireland and Israel.

Outsourcing in Russia has involved the participation of major US firms such as Sun Microsystems, IBM and Intel, each of which set up R&D centers in Russia, employing significant numbers of high-level programmers. Another form of collaboration has involved contract research between firms set up by the sprawling scientific-research institutes in Akademgorodok (Science City, with a heavy concentration of research and scientific establishments), Moscow, St. Petersburg and other major cities, with firms in Europe and the US. In addition to private sector participants from the West, the US Department of Commerce, as well as non-profit foundations, have invested in efforts aimed at gainful employment of scientists and engineers in Russia.[21] Under the impact of these influences and with the transition process having reached a mature stage, a dynamic software sector is now emerging in Russia, and is playing an increasing role in the global supply chain in software. Russian firms now participate actively in outsourcing related to applications programming and project consulting, as well as in turnkey assignments and basic research.

A survey by the authors of 48 Russian firms involved in outsourcing indicates that the sector is benefiting from the relatively large number of skilled programmers available at costs far below those in Europe or the United States[22]. Outsourcing has been a critical factor in the growth of the sector, with more than 50% of output exported to the US or Europe. The firms have built on the available, underemployed expertise. The training profile of the labor force in these firms indicates substantially more education than is found within the Indian labor force, as shown in Table 3-7. This aspect of the Russian advantage is perhaps the greatest competitive factor faced by Indian firms. One Indian firm, Mascon Information Technologies, has responded to this challenge by purchasing a Russian software manufacturer.[23]

---

[21] See the US Department of Commerce BISNIS web site for further information on these programs and outsourcing to Russia in general.

[22] The survey is part of ongoing research on the Russian software industry, which will be published separately.

[23] VARITA (2003).

Table 3-7. Educational Levels of Employees, Russian Software Outsourcing Firms
(With Rough Equivalence in Degrees; Percent of Total Employees)

	Russia	India
Bachelor of Technology	11.1%	42.9%
Master of Technology	49.5%	7.2%
PhD	8.5%	0.7%
MBA	1.0%	5.2%
Other Technical	22.3%	35.3%
Other	7.6%	8.7%
Source: Author survey, Fall 2002 and Dossani (2003).		

Table 3-8. Who Facilitated US Business Contacts with Russian Firms?

Type of Facilitator	Percent of Russian Firms
US Firm	75%
Russian Firm	80%
US Govt. Agency	2%
Russian Govt. Agency	0%
Russian Diaspora	35%
Source: Author survey, Fall 2002.	

Table 3-9. Assessment by Russian Software Outsourcing Firms of Competitive Strengths of
Rivals. (In Decreasing Order of Relevance and Importance, 1=highest to 5=lowest relevance).

	India	China
Lower Wages	1.8	1.9
Easy Availability of Labor	1.9	2.2
Telecom Infrastructure	2.8	2.5
Network of Research Institutes	2.8	2.5
Math and Science Background	4.0	3.0
Source: Author survey, Fall 2002.		

Lack of trade organizations, business skills, financing and marketing experience are seen as current impediments to growth. Most Russian firms now breaking into the outsourcing market are privately held, without direct government support (although many of the staff were trained in Russian research institutions or state enterprises). They rely primarily on private business contacts—either Russian or US firms—for most of their business development, as shown in Table 3-8. The role of the diaspora is increasing, both along the US-Russia as well as the Israel-Russia axis.

China shares some similarities but also has important differences compared to India or Russia as a place with long term potential for software industry development and outsourcing. (See Table 3-9 for a summary of the Russian software sector view of competitive differences of China and India as compared to Russia.) China has a large workforce with a level of mathematical/scientific training at least equal to India's and a similar price structure. The country's stock of research professionals has risen rapidly over the last few decades. China's institutional infrastructure for supporting services sectors is less well-developed than in India, and the country shares with Russia a lack of experience in business organization and marketing and some English language barriers. Despite these barriers, China appears to be moving towards becoming an important software site. As noted elsewhere in this chapter, the country is already seen as an important site for hardware producers, both because of low costs of production and because of the potential market. In addition, China is becoming a resource for Indian outsourcing companies, which can combine their business expertise with the skills available in China to provide more diversity in locations (in the face of political instability on the Indian subcontinent), a greater mix of labor force skills to their clients, and the synergy that comes from welding their software excellence to the hardware expertise of Chinese firms.[24]

The global expansion of software production through outsourcing centers on several continents creates further economic interdependencies and pressures for restructuring the industry within the US. Many outsourcing centers experienced slowdowns in demand with the economic dip in the US in 2001/2002. However, in some locations (most notably India) the slowdown was soon followed by an upsurge in demand as US firms tried to economize through moving additional functions to less expensive off-shore.[25] The role of high-tech services and software employment in California is likely to shift as these centers grow and mature. The question is whether the outsourcing will provide a cushion of lower cost labor to allow firms to continue operating economically in California and other parts of the US, or whether the new

---

[24] Lingblom (2002).
[25] Cole (2002), Thomas and Daga (2002).

agglomerations will become a direct competition to California firms in software and systems development.

3.10   FOREIGN COMPETITION

The 1997 interviews found that firms in all sectors of the industry (although not all firms) were concerned with foreign competition (as well as with competition from other California and US firms). For software firms, the main foreign competition faced was in foreign markets, rather than domestically. Much of the competition came from firms producing products more specialized for the needs of the foreign market in terms of standards or language. Producers of high-end software products faced competition both domestically and abroad from firms seeking to produce lower cost alternatives, with US firms being the major competitors in US markets. One strategy for dealing with this competition (as well as for expanding into related areas) was the acquisition or merger of competitor firms.

For component and PC manufacturers, foreign competition is much more significant in US markets as well as world wide, but a subset of firms in these subsectors saw US competition as more significant than foreign competition. For firms with strong market dominance (including some mainframe, specific component and network manufacturers), foreign competition might be unimportant for some products but significant for a subset of products. All computer cluster firms were concerned about US limits on the export of encryption technology and the opening given to foreign competitors in the worldwide market by these limitations. Software firms and some hardware firms were concerned about piracy and counterfeiting.

3.11   LOCATION DECISIONS AND POLICY ISSUES

Computer cluster firms constantly use location as a strategy in reaching new markets and in remaining competitive. They face the challenge of balancing the advantages of the combination of resources available only in locations such as California, with the high costs in the core areas where computer cluster firms locate. In the 1997 interviews, respondents indicated that the quality of California's workforce is key to the continued presence of computer-related firms in Silicon Valley and other locations throughout the state. All but one firm referred to skilled labor as a primary reason for remaining in California. The synergy provided by the proximity of many firms involved in computer-related activities also keeps many firms in the state. Specifically, firms mentioned the flow of information, proximity to investors, the flow of new ideas and the opportunity to acquire start-ups as advantages of being part of the California computer network. Related to this, communications infrastructure was seen by several firms as particularly strong in California. Other proximity advantages--to west coast markets and suppliers and overseas markets and suppliers--are also seen as advantages for a number of firms, as shown in Figure 3-10. The California quality of life was cited by about half of the firms interviewed.

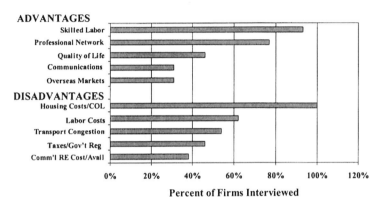

Figure 3-10. Advantages and Disadvantages of California Location to Computer Cluster Firms. Source: Responses to interviews with computer cluster firm executives, 1997.

Despite these advantages, firms have several concerns about a California location. Housing and other living costs, and as a result labor costs, were the most frequently mentioned disadvantages. Many firms felt the state income tax made labor cost problems more severe. While California labor was seen as skilled, it was not always easy to hire skilled labor in a competitive growth environment. About one-third of firms also mentioned commercial real estate cost and/or availability as problems with California locations. Beyond labor related issues, transportation congestion also troubled many firms. Government regulations, in the areas of workers compensation and land development, were seen as an issue by several firms, although a few also mentioned local and state programs to ease the effects of regulation as a plus.

Of firms expecting to expand, the majority were considering sites outside of California, either elsewhere in the US or overseas. They cited both "push" and "pull" reasons for this trend. California costs and regulations were the major push factors (although the "push" acted only to shift expansion outside of the area, and was not leading to relocation). Firms were drawn to out-of-state locations by land and labor costs, ease of transportation, location of foreign markets, and the many incentives offered by other states and nations.

The period following our interviews brought a series of challenges to computer cluster firms in California. During the 1990s, as the industry expanded rapidly, firms were faced with the decision of how to grow. A number of software and networking firms (including Siebel, Cisco, Sun, and Sybase), despite high labor costs, high housing costs, and worsening traffic congestion, adopted strategies of site expansion to satellite areas within the San Francisco Bay Area, close to their main operations and also to larger pools of housing.[26] Some firms also took advantage of this period of strong revenues to expand their international opportunities. Oracle, for example, followed opportunities in France, Thailand and Vietnam during 2000; Autodesk entered a joint venture with an Indian information technology firm to produce a new product.[27]

Once the contraction began in 2001, firms had to reduce costs. Layoffs were common. Some were spread widely, affecting foreign locations as much or more than the core employment in California. Other firms took the strategy of making heavy cuts in US manufacturing sites, consolidating manufacturing or services operations at lower cost overseas sites. Cutbacks of 15% at Silicon Graphics led the firm to close a plant in Switzerland in 2001. Quantum closed a Colorado manufacturing facility while investing in Malaysia in 2000. Advanced Micro Devices closed two chip plants in Texas but also cut its workforce in Malaysia in 2001.[28] The software firms that had planned expansions within the San Francisco Bay Area sharply cut back those plans, in some cases paying costly penalties for broken leases.[29]

3.12   POLICY CONCERNS

Interviews with firms indicated that they looked to state government for assistance with factors in California that affect competitiveness and to the federal government for assistance in areas directly related to foreign trade, imports and immigration. Government regulations and taxes were the most frequently cited state government programs that made it harder to remain competitive while located in California. Often it was the cumbersomeness of the process, rather than the absolute standard, that was most problematic for the firm. One firm gave the example of an overall environmental standard that could be met by the firm, but had become costly and cumbersome be-

---

[26] A whole series of articles in the *San Francisco Business Times* during 2000 document these plans.

[27] *San Francisco Business Times* (January, August and November 2000).

[28] Corporate financial reports published in Lexis-Nexis.

[29] Ginsberg (2002).

cause the firm must repeatedly file for permits related to that standard with each minor facility change.

Firms also mentioned a number of statewide programs that could help them in general with competitiveness, although not specifically in the international arena. Many mentioned the general need for changes in the regulatory process, especially in the areas of environmental controls, land development, and human resources. The need to maintain a strong education system was also emphasized. For example, one firm pointed to education as important not just as a means of producing the required skilled labor force, but also as an element in the recruitment of new employees, who seek good school districts for their families. Finally, a number of firms pointed to the incentives provided by other states and other nations as a significant factor in drawing expansion out of California. At least one firm had production facilities in a high cost European location chosen because of the generous concessions offered by the local government.

The major computer firms saw only a narrow scope for state action with respect to global operations of the firms. The state could help, primarily through trying to influence federal level policy, in the area of encryption controls and policy towards software piracy. These were also the areas of most concern at the US policy level. The firms universally felt that encryption controls were a significant limitation in the area of foreign sales. Most firms also sought protection for intellectual property rights abroad.

3.13   APPLICABILITY TO OTHER SECTORS

The computer cluster is among the most globally oriented sectors in California. Other California manufacturing sectors with high relative shares of trade include aerospace, instruments, and apparel. Food processing, beverages, agriculture and the motion picture industry are also significant exporters, although trade is a relatively low share of food processing and beverages income. The basic typology of stages of globalization is applicable to all of these industries, but the manifestations differ among the industries and from the computer cluster. All of these sectors have some degree of global presence beyond Stage I.

Some, like food processing and beverages, have many "Stage I" operations. Many food-processing companies interviewed for the 1997 study grew and processed their products within California and then exported from within the state. Foreign competition and trade agreements such as NAFTA have put cost pressures on other food processing companies, which are leading to moves towards imported "inputs" (e.g. asparagus grown in Mexico and distributed by fresh products companies in the US) and production overseas (e.g. Chilean and Italian wine production by California firms—Stage II or III).[30]

At the other extreme, high-tech manufacturing outside of computers, such as instruments manufacturing, can be expected to have globalization patterns similar to the computer cluster, especially with regard to imported inputs and specialized centers of manufacturing overseas.

3.14   CONCLUSIONS

Several key points emerge from examining the computer cluster:

1) Exports versus Foreign Sales: A growing market for products from California companies will not necessarily be met through export growth from California establishments. For the computer cluster, in an intensely competitive foreign environment, demand abroad is likely to be met by sales from foreign subsidiaries or affiliates of US and California multinational enterprises, with only a small proportion of the direct job growth (although perhaps a larger share of income growth) occurring within California.

2) Increasing globalization affects the employment mix in California's computer cluster, and could have similar effects on other sectors. As more production occurs overseas, the R&D activities and much of administration appear closely tied to California. An increasing proportion of California jobs in these sectors are high paying and technically demanding. Nevertheless, even here globalization plays a role. In the late 1990s and 2000, high demand and short supply of knowledge workers encouraged some firms to set up entire divisions of programmers overseas. In other sectors—food processing for example—competition from abroad may lead firms to seek other cost-cutting techniques, such as mechanization to reduce labor costs.

3) Imported intermediate inputs and overseas production have complex effects on California's economy. Global trade has the potential to strengthen California firms, providing inputs that help to maintain lower production costs and to diversify markets for products, thus cushioning fluctuations in demand. The ability to reduce costs through imported inputs and overseas production has a negative effect on job production levels in the state but also improves the competitiveness of firms against other producers.

4) Globalization may strengthen the advantages of service sector firms in California relative to manufacturing. In the computer cluster, for exam-

---

[30] Kroll and Kirschenbaum (1998) includes a case study of the food processing sector.

ple, the software firms maintained non-manufacturing activities primarily within the US, while moving manufacturing activities abroad.

5) Foreign trade is a factor leading to the restructuring of production and jobs in California. In the computer sector, the drop in manufacturing jobs during a period of sales expansion was counteracted by a rise in services jobs in computer-related sectors.

6) Environmental and land costs continue to play an important role in competitiveness of California firms in international markets and the long-term future of these firms in California. Even where there are strong advantages to the industry from a California location, the factors underlying these advantages (skilled labor, underlying educational institutions, fertile land, adequate water supplies) need continued attention if the sectors are to remain competitive.

The case studies suggest that many California firms have strengths and strategies that will allow them to compete in global markets, but that strong global exporters may also quickly become global producers. California economic development policy with regards to foreign trade must look at the topic in terms of its effects beyond the immediate sales levels in export areas to the broader impacts on firm structure, net job levels and occupational distribution. We return to this topic in Chapter 7.

Chapter 4

Foreign Outsourcing and Domestic Industry[1]

Chapter 3 in this book describes at length the role of high-tech firms in globalization. It addresses both issues of global production and trade, and the consequent restructuring challenges faced by these firms. In this chapter, we look at the phenomenon of foreign outsourcing and the impact it has on labor market inequality and value-addition in domestic industry.

In the United States, the wage gap between blue-collar (production) and white-collar (non-production) workers in the manufacturing sector has grown from approximately $7,500 in 1980, to nearly $24,000 in the year 2000. A similar and more acute process has been underway in the state of California where the wage gap has increased from around $10,000 to over $30,000 over the same period.[2] The US as a whole has lost close to three million manufacturing jobs in this period, with the bulk of the decrease coming from within the ranks of blue-collar workers.

Although this pattern of growing wage inequality has been well documented, its causes remain a matter of contention within the economics profession. Some attribute perhaps one-sixth to one-third of the increasing inequality to the growing forces of globalization—international trade and investment abroad, with the former competing away manufacturing jobs and the latter simply exporting them. Others find the globalization effect to be

---

[1] This chapter is an extension of work done with the collaboration of David Howe, a former colleague at the Fisher Center. The part of the chapter dealing with the period 1987-1992 basically reproduces, refines and updates the results of Bardhan and Howe (2001).

[2] Authors' calculations based on *Annual Survey of Manufacturers*.

negligible: they believe the phenomenon of growing inequality to be linked with technological change or perhaps exogenous shifts in preferences and product demands. There is little doubt however, that the changing landscape of manufacturing in the United States, the emergence of serious manufacturing competition abroad, the dominance of services in national output, as well a number of other economic and political factors have all contributed to this phenomenon.

Foreign outsourcing has gained momentum over the last few decades, particularly in the highly traded high-tech sectors.[3] Trade in intermediate products, particularly in the high-tech sectors, is one of the defining themes of this book. Indeed, one of the signal attributes of a manufactured high-tech product is the extensive nature of its value-chain, the number of intermediate products, services and manufacturing steps that go into the final output. Progress in transportation, communications, as well as in standardization has significantly increased the fragmented nature of production. The high-tech value-chain is now a multilateral production mosaic, spanning many countries. In some cases, inputs are outsourced from the US to arm's length suppliers, and in other cases the suppliers are affiliates of multinationals based in the United States, or, in the case of foreign-owned multinationals, the affiliate in the US outsources its inputs to another affiliate abroad, or to its headquarters at its home base. In chapter 5, we deal with the latter kind of trade in intermediate inputs – known as intra-firm trade.

In the first part of this chapter, we deal with two interrelated issues; we examine restructuring in manufacturing sectors in California in the context of foreign outsourcing. At one level, it contributes to the debate on inequality by looking exclusively at the economy of California, a state where wage disparities, and the forces of both global economic integration and technological change are at their most intense. In this context, we focus primarily on the impact that global linkages have on the economic fortunes of blue-collar and white-collar workers in the manufacturing sector, referred to as production and non-production workers, respectively.[4] To this purpose we

---

[3] Original electronic equipment manufacturers are one industry where over half of the US and Canadian firms outsource at least part of their production abroad; plastic molding is a popular candidate for such contract work (Elliott, 1998). Input suppliers themselves represent a highly organized sector, with firms managing facilities in multiple countries. That there can be considerable scope for substituting partly-assembled inputs for in-house labor is exemplified in the auto industry, where GM, Ford, Chrysler and Toyota purchase 53%, 62%, 66% and 75% from outside foreign and domestic suppliers, respectively (Taylor, 1994).

[4] These workers are sometimes called unskilled (blue-collar) and skilled (white-collar) workers. The general convention in the official data, the *Annual Survey of Manufacturers* (ASM) and the *Census of Manufacturers* (COM), is to refer to them as production (blue-collar) and non-production (white-collar) employees.

evaluate the impact of imports of intermediate inputs on the relative wages and employment of blue-collar workers.[5]

California is a region that sustained an acute recession in the early 1990s and experienced both divergent demands for blue- and white-collar workers and extensive exposure to foreign competition. Over our 1987-92 study period, the wage gap between production and non-production workers in Californian manufacturing establishments increased from $17,600 annually to $19,200, a change of $1,600 (in constant 1992 dollars). In our comparison region, comprised of the remainder of the US (or RUS), there was a smaller increase in manufacturing wage inequality, amounting to $1,200. Our results suggest that foreign outsourcing may account for a significant part of the increase in relative inequality between blue- and white-collar labor in California and the rest of the US. In terms of restructuring as a response to a recessionary environment, we find that industries with sharper sales declines were more likely to restructure by substituting imported intermediate inputs for domestic blue-collar labor.

In addition to the impact foreign outsourcing has on relative demand for unskilled labor, our approach, which incorporates imported inputs, allows us to simultaneously analyze the nature and timing of adjustment and restructuring undertaken by firms. Empirical work and a casual reading of the business press together suggest that industries reengineer their labor force in fits and starts, or to borrow a phrase from another discipline, in a series of punctuated equilibria: downsizing and restructuring is episodic, and often coincides with periods of low product demand and financial difficulties.[6] For economists, this is a puzzling phenomenon because it entails extended periods of non-optimizing behavior.

In the final part of the chapter we also examine the impact of foreign outsourcing on high value-added jobs. Looking at the phenomenon of restructuring through outsourcing from the viewpoint of a firm, it is clear that this cost-cutting exercise leads to higher value addition through lowering of input costs. Many observers have pointed out that over the years, the US as a whole, and California in particular, have managed to retain the higher value-added jobs. Here we show how that has come about in the context of foreign outsourcing, which has nibbled away at the lower end of the manufacturing spectrum in terms of value-addition per employee. The slow deindustrialization and loss of manufacturing jobs mentioned in the beginning of this chapter are partly a manifestation of this trend.

---

[5] Feenstra and Hanson (1996a) study this phenomenon of "foreign outsourcing", whereby higher imported inputs lead to lower relative demand for blue-collar labor.

[6] Of course restructuring can also occur during booms: Greising (1998) documents a number of companies that are downsizing even in the face of strong company earnings.

Our period of analysis spans the years 1987-1997, encompassing the three most recent Economic Censuses of 1987, 1992 and 1997, and hence includes the recession of the early 1990s that affected California so severely. The period from 1987 to 1992 deals with restructuring through foreign outsourcing during a recessionary downturn, and the impact that had on relative inequality between production and non-production workers, which increased during this period. The period 1992-1997 is qualitatively different for manufacturing in California. First, this is a period of growth, albeit of an uneven nature, and second, relative inequality *decreased* in these five years, as measured by our metric, which is the share of blue-collar wages in total payroll. Our emphasis in that period is therefore somewhat different, and deals mostly with the high-tech sectors, their imports of intermediate inputs and the subsequent impact on value-addition.

The story that unfolds in this chapter is that of increased foreign outsourcing, more actively resorted to by firms as an adjustment response during downturns, leading to greater inequality between skilled and unskilled workers, but which simultaneously increases value-addition per employee.

4.1   LITERATURE REVIEW

Standard international trade theory, embodied in the Stolper-Samuelson theorem and the Factor Price Equalization and Insensitivity (FPI) theorem, implies that changes in the international marketplace are communicated through relative price changes.[7] The FPI theorem contends that changes in factor prices, e.g. wage levels, are caused by changes in prices of traded products, and in the absence of the latter, simple quantitative changes in import levels should not affect wages. In contrast, in a study of the impact of import price indexes on wages and employment in nine US manufacturing industries, Grossman (1987) reports that the elasticity of domestic wages with respect to import price indexes is statistically indistinguishable from zero. Whether these results reflect an underlying insensitivity of wages to international factors or merely systematic bias in available price data remains unclear. At the very least, Krugman and others have noted that the period of greatest trade expansion—from 1950 to the mid 1970s—generally *preceded* the span of time with growing wage inequality.

The debate in the discipline centers on trade theorists, who argue in line with the FPI theorem that the labor markets will only be impacted by a change in prices of traded goods, and labor economists who maintain that imperfect measurement of product quality renders us incapable of perceiving such drops in final good prices and that regardless, trade in goods embodies

---

[7] See Caves, Frankel and Jones (2001).

the factors employed. Hence, an increase in the level of imports is tantamount to an increase in the supply of the embodied factor, and therefore a fall in its relative price, *i.e.*, that workers will be displaced or forced to take wage cuts. Labor economists therefore conclude that regressions relating wage changes to import levels are appropriate.

Our study utilizes a different approach, originally developed by Feenstra and Hanson (1996). We focus on one aspect of globalization, the substitution of unskilled in-house labor with foreign intermediate inputs. Feenstra and Hanson point out that such outsourcing tends to narrow the range of economic activities of a firm and hence changes the relative demand for different types of labor within the firm. A general equilibrium model in Baldwin and Cain (1994) shows this mechanism to be theoretically sustainable: imported inputs allow foreign labor to compete with unskilled domestic labor even in the absence of changes in output prices. The reduction in relative demand for blue-collar workers can thus be partly attributable to the increase in the share of imported inputs used by an industry.

Since our aim here is to simultaneously address the issue of restructuring through this kind of outsourcing, we also take inspiration from a different branch of literature. Identifying a period of restructuring—when production processes within an industry are reengineered while production employment declines—poses some problems for the researcher since it must be distinguished from the temporary shedding of labor that occurs during a downturn. Constructing numerical estimates of efficiency-augmenting innovations is also a challenge: measures of capital are problematic since it is difficult to distinguish between capital that substitutes for labor and capital that complements it.[8]

Davis, Haltiwanger and Schuh (1996) explore the link between restructuring and economic downturns. Using the Longitudinal Research Database, a dataset that measures expansions and contractions of employment at the plant level, the authors report mixed evidence that restructuring takes place predominantly during recessions. Not surprisingly, job destruction (or more accurately, plant-level quarterly declines in employment) is countercyclical: job creation falls and job destruction rises during a downturn, leading to a net job loss. Job destruction is persistent: most of the plant-level quarterly declines in employment continue into the next year, notwithstanding seasonal effects. The sum of job creation and destruction, termed "job reallocation" by the authors, is strongly countercyclical, indicating to them that the recessions have a connection with the restructuring process. At the same time, they also emphasize that most of the variation in job reallocation is

---

[8] Also, capital measures are derived from investment data and are sensitive to assumptions regarding depreciation. See Katz and Herman (1997).

associated with neither aggregate nor sectoral changes in demand, but rather with "idiosyncratic shocks", that is with conditions that are peculiar to each individual plant, suggesting that restructuring is also an ongoing process.

Caballero and Hammour (1994) explain job destruction's greater cyclicality with a model that assumes that costs of creating an additional productive unit (whether a plant or a process) rises with the number of productive units being created in the economy. As a result, variations in output cannot be accommodated purely by variations in job creation, and recessions serve to "cleanse" the economy of obsolete technologies.

Some evidence exists that both foreign competition and economic downturns promote structural change within manufacturing, although the two are typically not considered together. Baldwin and Caves (1998) survey the empirical literature showing that industries experiencing more intense foreign competition tend to show greater productive efficiency. In the same article, they use panel data to show greater industry turbulence, that is more plant closings, openings, company mergers and the like, in sectors more exposed to international trade flows.

Feenstra and Hanson (1996b) note that foreign outsourcing can represent a source of increasing income inequality in the U.S. to the extent that the labor embodied in the imported inputs substitutes for domestic unskilled labor. Although they did not establish such a relationship during the 1972-1979 interval, they did obtain statistically significant results for 1979-1990. They also note that their results were primarily connected with changes occurring during just two years, 1979 and 1981. Since partly manufactured inputs are typically not traded publicly, a reliable and accurate imported input price index would be difficult to construct. As an alternative, Feenstra and Hanson use changes in the volume of imported inputs as a proxy for changes in lower-skill input requirements, noting that their approach is consistent with the general equilibrium framework developed by Baldwin and Cain (1994) and others.

The 1980s witnessed an upsurge of popular articles in the press about the increasing phenomenon of outsourcing and the export of American jobs abroad. At the same time, a number of economists, such as Laura Tyson (1987) and others, started pointing out that although a number of blue-collar manufacturing jobs had migrated abroad, this process was also accompanied by the retention in the US of higher value-added jobs. The key to a competitive policy, it was maintained, was to nurture innovation and dynamism in the industrial sector that would keep creating higher value-added, higher paying jobs. In a more empirical context, Egger, Pfaffermayr and Wolfmayr-Schnitzer (2001) show that outsourcing of Austrian manufacturing to East European countries significantly improves domestic growth in total factor productivity.

Due to the paucity of data at the state level, as well as the difficulties in comparing data from 1997 with that of previous years due to the introduction of the new industrial classification system, we set ourselves a much more modest goal. Our analysis of the impact of outsourcing on value-addition, as well as more generally of the period 1992-1997, is therefore of a rather descriptive character. We cover the upturn in the high-tech industry in that period and describe the changes in wages and employment. We also follow the evolution of value-addition per employee in manufacturing as a whole and in high-tech sectors and relate it to the continuing increase in the share of imported inputs in manufacturing.

In the next two sections, we describe our empirical model and use it to consider whether the recession of the early 1990s produced a significant outsourcing effect, together with a concomitant rise in inequality between blue- and white-collar workers. In addition, we examine whether episodic restructuring can be demonstrated in a cross-sectional, cross-industrial context. We will see that the kind of fundamental changes in production reflected in foreign outsourcing occur predominantly, though not exclusively, in industries experiencing steeper declines in sales.

4.2 EMPIRICAL MODEL[9]

To attain the necessary variance in the regressors, we require data from a region that has experienced a substantial downturn in the recent past and has encountered both divergent demands for blue- and white-collar labor as well as exposure to foreign competition. In addition, the region must have sufficient economic activity to report data for a large sample of SIC codes. California fits these criteria very well. Of course, the whole of the US shares these characteristics as well, albeit to a lesser extent. We therefore include for comparison purposes the results from another region, the US apart from California, which we call the Rest of the US (or RUS).[10]

Data from the *Census of Manufactures* (COM), the *Annual Survey of Manufacturers* (ASM), the NBER productivity database, and the Department of Commerce's Trade File were used. We estimate a model based on Feenstra and Hanson (1996a and 1999). The following model is estimated:

---

[9] Sections 4.2, 4.3 and 4.4 are based on Bardhan and Howe (2001), with some corrections and changes.
[10] As the Californian economy is approximately one-seventh of the US, using the entire US as a comparison group would be problematic.

(1) $\Delta\varpi_i = c + \beta_1\Delta s_i + \beta_2\Delta(K_i/Y_i) + \beta_3 CYCLE_i + \beta_4 DEPTH_i + \beta_5 DEPTH_i * \Delta s_i$.

where,

(2) $\Delta s_i = \Delta(O_i/M_i)$

(3) $O_i = \Sigma_j (m_{ij}*(M_i/\Sigma_j m_{ij})*\{I_j/(Y_j - X_j + I_j)\})$,

Variables and symbols are defined as follows:

- $\Delta\varpi_i$ is the change in the share of blue-collar labor payroll over total payroll at the four digit SIC level. The variables are first differences between the levels in 1987 and 1992. These years correspond to Economic Census years, when detailed 4 digit data for California can be obtained from the Geographic Area Series.
- $\Delta(K_i/Y_i)$ is the change in the capital-sales ratio, used as a control variable in the regression. Capital data for the capital-sales ratio were gathered from the NBER productivity database for the years 1987-1991.[11]
- $CYCLE_i = (92 \text{ Sales}_i)/(\text{Maximum } (87 \text{ to } 92) \text{ of Sales}_i)$ is the ratio of industry sales in 1992 relative to the maximum value of industry sales in the period 1987 to 1992. This variable identifies the sector's position in 1992 relative to the business cycle of the period 1987 to 1992.
- $DEPTH_i = (\text{Minimum of Sales}_{i,} (87 \text{ to } 92))/(\text{Maximum of sales } (87 \text{ to year of minimum sales}))$ is the ratio of the minimum sales from 1987 to 1992 relative to the maximum sales between 1987 and the year of minimum sales. This variable reflects the intensity of the industry downturn.
- $\Delta s_i = \Delta(O_i/M_i)$ denotes the change in the share of imported inputs over total material inputs net of energy for the sector i.
- O_i is the total amount of imported inputs used in producing good i.
- m_{ij} are the material inputs from industry j used in the production process of industry i.
- M_i is the total material inputs, net of energy, used in producing sector i goods. The term $M_i/\Sigma_j m_{ij}$ is greater or equal to one and adjusts for the fact that not all materials are reported by type for different industries, and therefore accounts for missing materials.
- The term $\{I_j/(Y_j-X_j+I_j)\}$ is the import intensity index for industry j. I_j are the US imports of this input, X_j are the exports and Y_j the total sales of industry j. The index therefore is a measure of the share of imports in the total domestic market for good j.

---

[11] 1992 capital data were available only in preliminary form at the time of this writing.

The key assumption underlying equation (3) is that the ratio of imported inputs to total inputs of industry j goods equals the ratio of total imports to the total domestic market for industry j goods. For example, if imported integrated circuits equal 25% of the domestic market, then an industry using $1 billion of integrated circuits as inputs is assumed to import $250 million of this input.[12] The imported inputs are then summed over the entire range of identified inputs for each output industry i. The share of imported inputs in total material purchases for industry i (s_i) then equals O_i/M_i.

The variables in this model have been constructed with California data except for s_i and (K_i/Y_i). The problem for these two variables is that the Economic Census (regional) does not publish state-level input data. Instead, we have relied on 4-digit *US* data, implicitly assuming that the California industries parallel the US industries, at least in these dimensions.[13]

Our a priori expectation about the sign on β_1, our main coefficient of concern in equation 1, is negative, since that would imply that an increase in the share of imported inputs in manufacturing leads to a decrease in the share of blue-collar payroll. Since the right-hand side variables may be determined simultaneously by firms, estimates of the coefficients on imported inputs may be biased. The instrument, CHEPII, or the "change in the export propensity of inputting industries" is used:

$$(4) \quad \mathrm{CHEPII}_j = \frac{1}{\sum_i m_{ij}^{92}} \left\{ \sum_i m_{ij}^{92}(X_i^{92}/Y_i^{92}) - \sum_i m_{ij}^{92}(X_i^{87}/Y_i^{87}) \right\}$$

Our instrument differs from our imported inputs measure in that it does not reflect changes in the input mix occurring between 1987 and 1992 and it uses an export share measure rather than the import intensity index. The instrument will be correlated with our imported inputs measure, since there is typically a positive correlation between an (inputting) industry's imports and its exports.[14] Also, holding the material weights constant allows the instrument to be independent of possibly endogenous shifts in the production process. That is, the instrument should be uncorrelated with the error term. The interaction term in (1) uses the instrument CHEPII*DEPTH.

---

[12] The basic equation is comparable to Feenstra and Hanson (1996b), as verified by private correspondence, except that our import intensity measure nets out the effect of exports in the denominator to capture the size of the domestic market more accurately.

[13] The *Census of Manufacturers* provides 4 digit intermediate input data. Rigby and Essletzbichler (1997) report, however, that input coefficients vary significantly by region at the 2 digit level. We minimize this problem by focusing on a substantially narrower SIC coding and considering changes rather than levels.

[14] Recent literature on intra-industry trade, even at the 4-digit SIC level, demonstrates the high correlations between exports and imports within an industry.

4.3   REGRESSION RESULTS

Table 4-1 shows summary statistics for the key variables. Over the period, California manufacturing production workers' share of total payroll fell from 45% to 41.5%, while in the rest of the US it decreased from 58.5% to 57%. In the key high-tech sectors for computers and semiconductors (SIC 357 and 367), the decrease was comparable, although the shares were much lower to begin with: 27.7% for California and 35% for RUS.

Table 4-1. Summary Statistics, Manufacturing Sector, 1987 to 1992

Variable	Region	Mean	Median	Standard Deviation
Change: Production Payroll divided by Total Payroll ($\Delta\varpi$)	California	-0.035	-0.030	0.061
	Rest of US	-0.016	-0.018	0.029
Change: Imported Inputs divided by Total Materials (Δs)	US	0.021	0.013	0.038
Cyclic Variable (DEPTH)	California	0.820	0.837	0.116
	Rest of US	0.883	0.898	0.079
Source: Bardhan and Howe (2001), US Economic Census 1987 and 1992				

Table 4-2 presents our regressions, which are estimated on a cross-sectional sample of 218 manufacturing sectors at the 4-digit SIC code level. The dependent variable is the change in the payroll share of blue-collar workers. The first column suggests a modest negative effect of imported inputs on the relative demand for blue-collar labor in California. In the rest of the US, column 4 shows a smaller and less significant coefficient. In the initial OLS specification in Table 4-2, the signs on the cyclic variables are highly significant for both California and the rest of the US in an unexpected direction: it would appear that cyclical downturns favor blue-collar workers in manufacturing at the expense of their white-collar brethren.

The F-tests show that the imported input and interaction coefficients are jointly significant for both regions. At the bottom of the table, we evaluate the overall magnitude of our particular form of restructuring with R:

$$(5) \quad R = \frac{d\Delta\omega}{d\Delta s_i} = \beta_1 + \beta_5 * \text{DEPTH}$$

For California, R is negative and significant for most values of DEPTH, i.e. over a range of one standard deviation on both sides of the mean. For RUS, the findings suggest a negative relationship between imported inputs and the relative demand for blue-collar labor, albeit to a slightly lesser degree and appears confined to sectors experiencing sharper downturns, since it is not significant when assessed at mean plus one standard deviation.

Table 4-2. Regression Results: Dependent Variable is Δϖ, Change in the Share of Blue-Collar Payroll

Variable	California (CA) Results			Rest of US (RUS) Results		
	OLS	OLS	Instrumental	OLS	OLS	Instrumental
Constant	6.6E-4	1.1E-3	4.2E-3	3.7E-4	4.6E-4	2.5E-4
	(2.7E-3)	(2.3E-3)	(2.8E-3)	(9.3E-4)	(9.4E-4)	(1.1E-3)
ΔImported Inputs Share, Δs, (β1)	-.151**	-3.79**	-5.35**	-.029*	-1.39**	-2.88**
	(.065)	(.419)	(.569)	(.015)	(.43)	(.72)
ΔCapital Sales ratio, Δ(K/Y), (β2)	-.026	.004	-.007	-.024	-.034**	-.039**
	(.014)	(.046)	(.045)	(.017)	(.017)	(.018)
CYCLE, (β3)	-.022**	.001	-.028	-.020**	.047	.053
	(.006)	(.062)	(.067)	(.003)	(.040)	(.052)
DEPTH, (β4)		-.02	.030		-.070	-.075
		(.064)	(.069)		(.043)	(.059)
DEPTH*Δs, (β5)		3.90**	5.36**		1.38**	2.60**
		(.450)	(.575)		(.451)	(.843)
F test (β1=β5=0)		43.7**			4.6**	
Adj R²	.41	.64	.61	.39	.35	.36
N	218	218	218	217	217	217

Addenda: Evaluation of Interaction Term: d(Δ Blue Collar payroll share)/d(Δ Imported Inputs) = β1 + DEPTH*β5						
Evaluated at	California			Rest of US		
Mean(DEPTH) – stdev(DEPTH)	.703	-1.051**	-1.579**	.803	-.280**	-.79*
Mean(DEPTH)	.820	-.598**	-.956**	.883	-.171**	-.58*
Mean(DEPTH) + stdev(DEPTH)	.936	-.146**	-.334**	.962	-.006	-.038

(Standard errors in parentheses), **Significant at 5%, * at 10%; Standard Deviation (Depth), CA = 0.116, RUS = 0.079.
Source: Based on Bardhan and Howe (2001).

Table 4-3. Share of Industries with Declines in Sales at Varying Levels

California

Yr	% Change GSP¹	Decline in Sales					
		>2%	>5%	>10%	>15%	>20%	>30%
1988-92	–	98.6%	90.8%	70.6%	53.2%	32.6%	16.5%
1989	4.7%	54.4%	38.3%	17.6%	6.0%	1.0%	0.0%
1990	2.8%	70.5%	58.1%	24.0%	17.1%	1.9%	1.0%
1991	-1.8%	90.4%	83.4%	59.0%	45.6%	26.3%	7.9%
1992	-0.5%	75.6%	65.5%	51.6%	40.6%	28.1%	10.9%

Propensity to restructure: d(D Blue Collar payroll share)/d(D Imported Inputs)

DEPTH	0.98	0.95	0.9	0.85	0.8	0.7	
OLS Estimates	0.027	-0.09	0.28**	0.48**	0.67**	1.06**	
Instrumental Estimate	-0.097	0.26**	0.52**	0.79**	1.06**	1.59**	

RUS

Yr	% Change GRP¹	Decline in Sales					
		>2%	>5%	>10%	>15%	>20%	>30%
1988-92	–	92.2%	72.5%	51.4%	32.6%	20.6%	0.5%
1989	2.2%	45.7%	22.6%	8.3%	1.8%	1.9%	0.0%
1990	-0.4%	58.1%	42.0%	10.7%	8.8%	1.4%	0.0%
1991	0.4%	89.9%	71.5%	50.3%	30.0%	17.6%	0.5%
1992	4.2%	72.9%	53.5%	38.3%	20.3%	10.2%	0.5%

DEPTH	0.98	0.95	0.9	0.85	0.8	0.7	
OLS Estimates	0.063	0.012	-0.15*	-0.19**	0.285**	–	
Instrumental Estimate	0.115	0.017	-0.23*	-0.61**	0.805**	–	

¹GSP: Gross State Product; GRP: Gross Regional Product.

* Significant at 10%; ** Significant at 5% level.

Source: Based on Bardhan and Howe (2001). The sales declines are from the Economic Census and Annual Survey of Manufacturing. The bottom part is calculated based on values of coefficients from Table 2 and the addenda in the bottom part of Table 2.

How the intensity of the imported input effect varies with the hardship experienced by the sector is indicated by the cross-partial derivative,

(6) $$\frac{\partial^2 (\Delta \text{blue-collar payroll share})}{\partial (\text{imported inputs}) \, \partial (\text{DEPTH})} = \beta_s.$$

which is positive and highly significant in both regions and specifications.

Table 4-3 shows how foreign outsourcing varies among industries experiencing milder and harsher declines in sales. The top row is derived from the DEPTH variable, and reports the maximum sales decline during the period of study. Since we were not considering restructuring that might have taken place prior to 1987, sales declines are measured only relative to peak years occurring during the 1987-1992 interval. The next set of rows show the share of industries suffering various declines in sales during selected years. Declines of 2% to 5% are not unusual, even during years of economic expansion such as 1989. Sales declines of 20% or more can be characterized as recessionary. Our model shows that restructuring in California is four to seven and one-half times as intense for industries experiencing a 20% decline compared with those whose sales dipped 5% (Compare the figure – 0.675 with –0.090, and the figure –1.061 with –0.257). In RUS, a 5% sales decline does not prompt the sort of restructuring that lowers relative demand for blue-collar workers. To the extent that such restructuring occurred in RUS over the period, it was two to four times as intense in industries whose sales declined by 20%, when compared with those experiencing a 10% decline.

To gauge the overall effect of imported inputs on our measure of the relative demand for blue-collar labor, R (from equation 5) can be evaluated at the mean value of DEPTH. Using those values from Table 4-2 and applying them to the change in imported inputs over the period, from Table 4-1, we can get the share of the change in blue-collar payroll ratio that is due to foreign outsourcing. In California, OLS and instrumental variables estimation implies that the increase in imported inputs accounts for one-third to one-half of the decrease in blue-collar payroll shares, respectively. In RUS, comparable calculations suggest that foreign outsourcing is responsible for about 20% to 50% of the relative increase in inequality between blue-collar and white-collar workers. However, the higher figure is not statistically significant.

Several explanations may account for the relative disparity between California and RUS, as well as the somewhat weaker results in the latter. Part of the explanation is compositional: the ten sectors with the largest changes in imported inputs accounted for 10% of California blue-collar employment in 1987, but only 5% of RUS.

The regressions also do not reflect the migration of Californian production facilities to lower wage states such as Arizona or Texas during the period. This problem was especially acute for SIC codes 3674 and 3679, semiconductors and miscellaneous electronic components, where blue-collar payroll shares declined for California but rose for the remainder of the US.

The 1987-1992 segment of Table 4-4 indicates some shifting of blue-collar jobs from California to RUS, at least for these two industries. The blue-collar wage gap between California and the rest of the US narrows and California blue-collar employment declines whereas RUS has the opposite experience. Yet, as the table shows, the migration of production workers does not fully account for the employment changes in the 2 regions, suggesting a more elaborate story than one reflecting a simple transfer of jobs from one part of the country to another.

Table 4-4. Employment Profile of Semiconductors and Electronic Component Industries

	Semiconductors		Electronic Components	
	California	RUS	California	RUS
Change in share of Blue-Collar Payroll 1987 to 1992	-0.033	0.019	-0.098	0.028
Change in Blue-Collar Employment 1987-1992	-3,100	400	-800	12,700
Change in Blue-Collar Employment 1992-1997	2,285	8,900	11,900	44,000
All In 1992 Dollars	California - RUS		California - RUS	
Blue-Collar Annual Wage Gap for 1987	$2,537		$2,716	
Blue-Collar Annual Wage Gap for 1992	$1,883		$2,305	
Blue-Collar Annual Wage Gap for 1997	$2,850		$6,291	
Source: Based on Bardhan and Howe (2001), Economic Census of 1987, 1992, and 1997.				

4.4   THE RESULTS EXPLAINED

Our results indicate that instead of indulging in continuous optimization behavior, firms tend to optimize in punctuated equilibria, or in fits and starts, and that the periods of intense restructuring coincide with recessionary downturns. Managers in troubled industries are more likely to adjust their production processes by substituting imported inputs for less skilled labor. This raises the suspicion that less vulnerable industries may be passing over efficiency-augmenting opportunities, at least those that involve workforce reductions. Although a casual reader of the business press might find this unsurprising—firms are often characterized as being "forced" to downsize—the theoretical basis for postponing cost cutting until reduced sales force the issue remains unclear. Below we discuss four sets of explanations: lumpy adjustment costs, dynamic considerations, agency explanations and labor relations.

4.4.1   Lumpy adjustment costs

A firm with only a small inefficiency in its production process may delay restructuring this process in the presence of lumpy adjustment costs. After all, a firm will bear the large fixed investment cost to upgrade its production process only at that point in time when the benefits of restructuring exceed the costs of restructuring.

4.4.2   Dynamic Considerations

Caballero and Hammour (1994) build a dynamic model of "creative destruction" whose main features appear to be broadly consistent with our results. In their model, technological progress is monotonic, with the newest production units embodying the most productive technology. Because entry is costly, production units with varying productivities can coexist. Technological change is determined by the exogenous rate of technical advance as well as two endogenous factors. These are the rate of entry of high productivity units and the varying age of obsolescence of the least productive and by assumption oldest units. A drop in output demand reduces the creation rate at the high productivity margin and increases the destruction rate at the low margin. One of the main results of the model is that destruction of the older units will vary more than the creation of new ones across the business cycle.

4.4.3   Agency Issues

It is not surprising that hard times may promote restructuring if credit constraints and negative cash flow create the possibility of bankruptcy. In less apocalyptic circumstances in contrast, shareholders may exert less pressure on industries suffering from larger sales declines, since industry-wide changes might be interpreted as reflecting factors beyond the control of management. At the same time, managers who favor a quiet life may defer restructuring that entails job loss, contrary to the interests of their shareholders.

There is also evidence for loss aversion and poor information acquisition among proxy voters: DeAngelo and DeAngelo (1989) observe that dissidents tend to cite simple measures of performance, such as negative accounting profits or stock declines from all time highs. Complex statistical analyses are invariably absent, the thinking being that holders of shares traded in liquid markets are unlikely to evaluate management in a meticulous fashion.

Of course, proxy contests are not the sole mechanism for changes in management. Warner, Watts and Wruck (1988) note that lagging stock returns can promote managerial turnover. The effect appears driven predominantly by the top and bottom deciles of the stock performance distribution. Furthermore, firms in poorly performing (2 digit) industries appear to experience higher CEO turnover, even after controlling for own-firm stock price effects. Whatever the reasons, managers in industries suffering from higher sales declines may feel extra pressure to cut costs.

4.4.4   Labor Arrangements

Work arrangements in traditional unionized or union-influenced settings typically allow for layoffs during economic downturns, subject to seniority practices. An implicit agreement seems to be at work here: a reluctant acceptance of layoffs during really bad times in exchange for security during marginally bad ones. This would tend to explain the sudden jump in propensity to restructure with deeper downturns. Workers operating under a salaried model of employment may receive greater assurances of job security, although this, too, is limited by the economic environment: job loss may be acceptable provided the firm is perceived as making good faith attempts to avoid such an eventuality. Under either structure, managers may believe that dismissals during economic expansions could hurt employee morale, leading to low effort and bad labor relations in general.[15] If this is the case, terminat-

---

[15] See Osterman (1993) for a discussion of the industrial and salaried models of employment.

ing workers during hard times is more acceptable since it is perceived as necessary.

Evidence for such a view is limited though: most studies of fairness in the economics literature have focused on wages and prices. A survey of white-collar survivors of restructured workplaces revealed that those who anticipated further layoffs were more apt to report lower loyalty and work effort as well as higher quit intentions.[16] Even if the latter attributes of employees are misperceived there is the possibility of an equilibrium of mistrust created on both sides, and the managerial response might be to preempt that.

4.5   THE UPTURN IN HIGH-TECH INDUSTRY AND CALIFORNIA ECONOMY, 1992-1997

The period 1987-1992 saw an overall decrease in the share of production workers' wages in total payroll in manufacturing as a whole, and an increase in the share of imported inputs in all inputs used in manufacturing. The previous sections demonstrated how the increase in foreign outsourcing and rising inequality between blue- and white-collar workers are linked. We also showed how foreign outsourcing is used as an adjustment strategy by firms during recessionary downturns.

In contrast, the period 1992-1997 was marked by a significant improvement of economic conditions for California in general, and for the high-tech industry in particular. The 1992-1997 segment of Table 4-4 indicates the reversal in fortunes for California's high-tech sectors:

- The wage gap between California (CA) and the rest of the US (RUS) again increases.
- Employment gains for RUS are much higher than in the earlier period, although this time it is not at the expense of California, since blue-collar employment in the latter also increases significantly.

At the same time, the share of imported inputs in all input usage in manufacturing has gone up steadily from around 10.5% in 1987 to 12.7% in 1992 to over 16% in 1997. In the high-tech sectors, the proportion was significantly higher to begin with and has increased further, as shown in Figure 4-1. Imported inputs occupy a significant share of overall imports into the US as well. As shown in Chapter 5, approximately 38% of all goods and merchandise imports of the US in 1992, were intermediate goods used in manufacturing. The figure increased marginally to around 39% in 1997. The im-

---

[16] Turnvey and Feldman (1998).

portance and significance of imports in general, and imported inputs and high-tech imported inputs in particular, is underscored by the fact that in 1997, high-tech imported inputs accounted for 25% of all imported inputs.

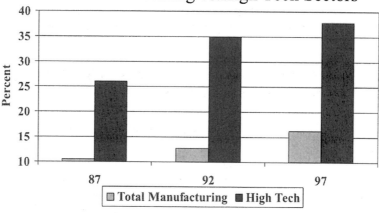

Share of Imported Inputs in Total Inputs:
Total Manufacturing vs. High Tech Sectors

Figure 4-1. Share of Imported Inputs in Total Inputs, Total Manufacturing vs. High-Tech Sectors. Source: See Appendix Chapter 5.

Together with a cyclical upturn in the economy after the just concluded recession of the early 1990s, the period 1992-1997 also saw a major structural and technological change in the economy with the coming of age of the Internet and of E-Commerce, as well as related sectors, such as wireless communications. Between 1992 and 1997 there was a significant increase in the share of the payroll going to blue-collar workers, both in California, as well as the US as a whole, while the share of imported inputs in total inputs increased further. The absence of a downturn, as well the increase in the share of blue-collar payroll share meant that a study similar to the one in previous sections was not possible. In addition, the Economic Census data for 1992 and 1997 are comparable only to some extent, and the imperfect matching of manufacturing sectors in the two years creates severe difficulties in the kind of econometric work that was done for the previous five-year period.

Figure 4-2 shows how the share of production workers' wages in total payroll has evolved over time. There is a significant drop between 1987 and 1992, particularly for California, but by 1997 it is already higher than a decade before. Since then it seems to have remained steady, although in all likelihood the present ongoing downturn, as of 2002, is also taking its toll on it.

The cyclical nature of this data might seem to pose a problem at first sight, since a one-time permanent restructuring was part of the initial working hypothesis. The manufacturing jobs which leave the shores of the US or California were thought to have gone for good. However, conceivably the latter explanation may still hold, since our data does not throw light on whether the increase in demand comes from new sub-sectors in manufacturing, which grew rapidly in those boom years. If that is indeed the case, then as with any new branch of manufacturing first set up in the US, an initial boost to blue-collar ranks would naturally follow suit, before subsequent development of skills and infrastructure abroad in less costly economic environments, followed by foreign outsourcing.

To gain additional insights it is worthwhile to look at some high-tech sectors that have been at the forefront of the restructuring process. We see from Figure 4-3 that for the high-tech sectors, which include computer, peripherals, components and semiconductor manufacturing, the share behaved in a fashion similar to that for manufacturing as a whole, through 1997. However, the shares of blue-collar payroll in California, and in high-tech sectors, are lower than the corresponding shares in the US and in total manufacturing respectively, and the differences at the level of individual wages are even more stark. By 1997, in California the wages of a blue-collar worker in the high-tech industry were $33,000, about half those of a white-collar worker in the same industry---$65,000, while in the US as a whole the corresponding figures were $29,000 and $56,000.

Thus, in spite of the cyclical nature of the data, there is evidence that the process of decreasing demand for blue-collar labor through foreign outsourcing may very well be continuing in the present downturn in the high-tech manufacturing sector. Apart from the anecdotal evidence in favor of this hypothesis, as well as the findings of the case study chapter in Chapter 3 of this book, one should also take into consideration the complex motivations for fragmented production in high-tech. Chapter 5 will shed some light on the role played by multinational corporations in trade in intermediate inputs. Within firm trade, or intra-firm trade in general, and intra-firm trade in intermediate inputs in particular, are determined by a host of additional factors, rather than merely being the outcome of foreign outsourcing brought about by cost-cutting due to local economic conditions. Tax considerations, transfer-pricing issues, global market potential and a number of other factors play a role. It is the cutting edge, innovative nature of high-tech in California in particular, that periodically boosts blue-collar demand, whether in pilot plants or initial manufacturing setups in the US, to be followed later by foreign outsourcing.

Share of Production Workers Wages in Total Payroll: California vs. USA

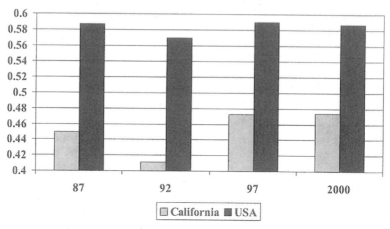

Figure 4-2. Share of Production Workers Wages in Total Payroll, California (CA) vs. US. Source: *Annual Survey of Manufacturers* and the *Census of Manufacturers.*

Share of Production Workers Wages in Total Payroll in High Tech Sectors: California vs. USA

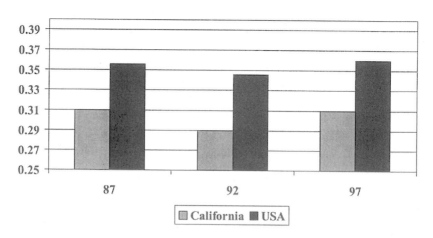

Figure 4-3. Share of Production Workers Wages in Total Payroll in High-tech Sectors: CA vs. US. Source: *Annual Survey of Manufacturers* and *the Census of Manufactures.*

4.6   FOREIGN OUTSOURCING AND VALUE-ADDITION

Apart from its impact on labor market inequality, foreign outsourcing affects firms in yet another way. The process of outsourcing inputs from abroad leads to lower costs, and hence for a given level of sales or shipment values, to an immediate improvement in value addition, since the latter is nothing but the difference between sales and material input costs. Of course, value addition increases from a host of other productivity enhancing factors, technological improvements and managerial innovations.

Figures 4-4 and 4-5 show the rapid increase in value addition per employee, both in manufacturing as a whole and in the high-tech industry in particular. Although the figures are in nominal dollars, the increase in value addition per employee for California during the period 1987-2000 is 108%, while for the US it is 82%. It is interesting that nominal wages of all manufacturing employees in the same period rose by 44% in California and 53% in the US as a whole. The increasing share of corporate profits in value-addition that the data implies, and for which there is some evidence in the national accounts as well, is part of the story behind the remarkable run up of stock prices practically throughout this period.

How much of a role in all these phenomena did foreign outsourcing play? To get an indicative answer we turn now to a simple regression analysis that will help us with an assessment of the contribution of foreign outsourcing to value-addition per employee. We estimate the following equation, similar to our earlier specification for the period 1987-1992. Unlike the earlier equation, the present model is in levels for the year 1997 and not in first differences, since the data for 1992 and 1997 are not comparable at a 4-digit level of detail. The other difference of course is that here our dependent variable is value addition per employee in each 6 digit NAICS code, v_i.

(7) $v_i = \text{constant} + \beta_1 s_i + \beta_2 K_i / Y_i$.

The right hand side contains both the imported input variable, normalized by sales and control variables such as capital sales ratio, from our earlier data set. We carry out OLS as well as estimation with instruments because of the simultaneity issue (see Table 4-5). The instrument used is the same as shown in equation (4). Most of the signs on the coefficients are in the expected direction, except for the OLS specification for the US as a whole. Although the fit of the models is not particularly high, and there are some concerns with the specification used, they do provide some indicative evidence of the positive impact of foreign outsourcing on value addition.

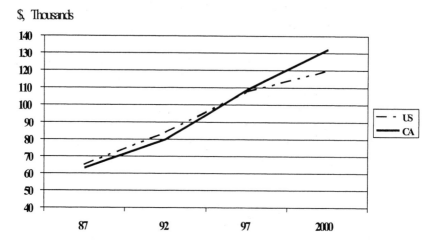

Figure 4-4. US and California (CA) Total Manufacturing: Value-Addition Per Employee.
Source: Authors' calculations based on *Annual Survey of Manufacturers* and the *Census of Manufacturers*.

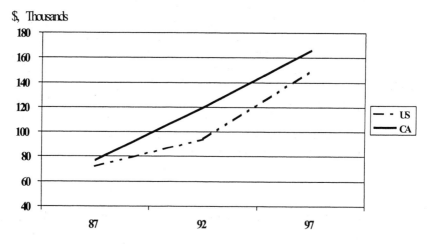

Figure 4-5. US and California (CA) High-Tech Manufacturing: Value-Addition Per Employee. Source: Authors' calculations based on *Annual Survey of Manufacturers* and the *Census of Manufacturers*.

Table 4-5. Regression Results, Value Addition Per Employee

Dependent Variable is Value Addition Per Employee	California		US	
	OLS	IV	OLS	IV
Constant	0.21*	0.11*	0.08*	0.06*
Imported Inputs/Sales (s)	0.35**	0.41**	0.25**	0.13
Capital/Sales Ratio (K/Y)	0.05	0.12	-0.14**	0.09**
Adjusted R^2	0.12	0.11	0.15	0.08

Note: * Significant at 5%; ** Significant at 10%; N = 330

4.7   CONCLUDING REMARKS

Our results for the years 1987-1992 confirm that trade plays a role in rising labor market inequality. Where California is concerned, we find that the forces of globalization, manifested in foreign outsourcing, particularly between 1987 and 1992, account for 33% to 50% of the increase in relative inequality between blue- and white-collar workers, where relative inequality is expressed as the share of the former in total payroll. In the case of the rest of the US (RUS), foreign outsourcing contributes 20% to 50% of the increase in inequality. At the same time, industries experiencing sharper sales declines were more likely to restructure their production processes by substituting manufactured imported intermediate inputs for domestic blue-collar labor. In California, a 20% sales decline during the period prompted a shift away from blue-collar labor that was four to seven and one-half times as intense for a given change in the share of materials that were imported, when compared with a 5% sales decline. In the rest of the US, 5% declines generated no perceptible restructuring, while there was some evidence indicating a significant effect for a 20% decline. This cross-sectional result is broadly consistent with Feenstra and Hanson (1996), who observed that foreign outsourcing was concentrated during the recession years of 1979 and 1981.

Why hard times would promote restructuring, separately from the usual laying off of production workers that occurs during downturns, is less than entirely clear. Lumpy adjustment costs provide a partial yet not completely satisfactory explanation for this phenomenon. Accelerated retiring of the least productive units during downturns and dynamic entry-exit issues provide another possible explanation. Furthermore, there is some evidence that poorly performing industries encounter greater shareholder pressure. The combined stance of employees who perceive downsizing as a recurring event—as opposed to a response to particularly hard times—and managers who internalize this perception on the part of employees may well set themselves up for a mutually reinforcing situation of mistrust. Restructuring of this nature, which has a propensity to occur in those industries undergoing exceptionally bad times, may be the managerial response to pre-empt bad labor relations.

Although the share of blue-collar wages in total payroll increased in the period 1992-1997, our contention that the structural shift is a secular change rather than a cyclical one still holds. The blue-collar jobs in manufacturing that were "exported" abroad have gone for good. It is unlikely that the "same demand" (read "same jobs") for blue-collar labor is back. This period witnessed extraordinary dynamism and innovation in domestic industry and in all probability it is the new plants, with new technology that are adding to the demand for production labor, and creating new kinds of blue-collar jobs. With finer detail in our data and greater comparability of the Economic Census of 1997 with 1992, it might have been possible to discern this pattern. The trend with imported inputs, on the other hand is much more straightforward, with steady increases throughout the entire period 1987-1997 and beyond.

Value addition generally, as well as on a per employee basis, has seen a significant rise. Although a larger proportion of the benefits have accrued to firms, in the form of higher profits, at least some of the employees—particularly the non-production workers—have gained as well. A case can be made that foreign outsourcing operating through the medium of higher value addition plays some role in these increased benefits to the economy.

Restructuring during a recessionary downturn through foreign outsourcing may be an even wider phenomenon than we initially surmised. Large segments of the business services sector, long thought to be among the most non-tradable part of the economy are now being outsourced to foreign countries, although the services industry is much more institutionally and culturally constrained than manufacturing. Business process outsourcing, in the form of call centers, accounts receivable/payable management, billing and forms processing, among other activities, is rapidly gaining pace in the current recessionary environment and we expect the phenomenon to accelerate in the coming years.

In Chapter 5 we analyze a particular kind of foreign outsourcing – to wit, that which takes place across borders, but within a transnational firm; this kind of trade known as intra-firm trade is of great importance in international trade. In that chapter we attempt, perhaps for the first time with US data, a study of intra-firm trade in intermediate inputs.

Chapter 5

Intra-Firm Trade and Intermediate Inputs
The New Stage of Globalization

Chapter 4 dealt with issues relating to the impact that foreign outsourcing has on domestic industry. We saw that firms resorted to foreign outsourcing as an integral part of a restructuring strategy during a recessionary downturn, and that this activity leads to an increased inequality between blue- and white-collar workers, while increasing value-addition per employee. As described in both Chapters 3 and 4, foreign outsourcing can be carried out by US multinational firms through arm's length suppliers based abroad, or through their own affiliates and subsidiaries. In addition, US affiliates of foreign multinationals can outsource to their headquarters abroad, or to subsidiaries and affiliates in third countries. Such intra-firm (within firm) trade in intermediate inputs is the object of study in this chapter.

5.1   INTRA-FIRM TRADE AND INTERMEDIATE INPUTS

In advanced industrialized economies, globalization has included key roles for both foreign outsourcing of intermediate inputs and intra-firm trade.[1] While both subjects have been studied by economists, their interaction and possible intersection (viz. transnational intra-firm trade in interme-

---

[1] See Markusen (1995), Irwin (1996), and Feenstra (1998) for surveys of the basic patterns for international trade with particular attention to intra-firm trade and international outsourcing.

diate inputs) have received relatively little attention.[2] Low-cost foreign outsourcing has long attracted many firms, whether part of a multinational enterprise or independently. Increasingly however, organizational considerations have motivated firms to use imported inputs from foreign affiliates, instead of inputs from arm's length domestic manufacturers; this change amounts to vertical integration across borders.

The computer industry is a particularly telling example of intermediate inputs and intra-firm trade. The complexity and sophistication of the end products of this sector allow specialized production activities and stages, and hence a large number of intermediate inputs. The geographical spread of the production base of most of the large multinational firms in this industry results in a brisk international trade in intermediate products.

The computer cluster case study in Chapter 3 confirmed the extent of foreign outsourcing and intra-firm trade in the high-tech firms of California. Indeed, signal attributes of a manufactured high-tech product include the extensive nature of its value-chain, the number of intermediate products, and the global, fragmented, nature of the final output. Progress in transportation, communications, as well as in standardization has significantly increased the fragmented nature of production, not to mention organizational and tax imperatives facing multinational firms. The high-tech value-chain is now a multilateral, multinational, production mosaic, spanning many countries and production locales, but often involving just one firm.

In this chapter, we study intra-firm trade and imported intermediate inputs, with special focus on the high-tech, computer, industry:

1) What are the determinants of imported intermediate inputs across shipping, or exporting countries?
2) What is the relationship between imported intermediate inputs and intrafirm imports?
3) Are there differences in the structure of intra-firm trade along industry lines?

Foreign affiliates of US multinational enterprises (MNEs) may provide either distribution or production facilities for their parent companies. In this chapter, we will focus on affiliates that function as production centers rather than distribution outlets for goods produced at the home location. Sales by these affiliates can be directed to a variety of customers:

---

[2] Recent studies of trade and sales by multinational firms and their affiliates include Brainard (1997), Zeile (1997), Markusen and Maskus (2001), and OECD (2002). Recent studies of international outsourcing include Hummels, Raporport, and Yi (1997), Campa and Goldberg (1997), Feenstra and Hanson (1999), and Swenson (2000).

1) To the MNE parent;
2) To other customers in the home country of the MNE;
3) To worldwide customers, not in the home country of the MNE.

Category (1) represents one form of intra-firm imports for the US. The other, parallel, form of intra-firm imports occurs when a foreign-based MNE ships goods to its US-based affiliate. Both forms of intra-firm trade are influenced by industrial organization issues such as transactions costs, as well as specific international trade factors such as tariffs and long-distance transportation costs. Strategic incentives for intra-firm trade are also found in tax related issues, exchange rate hedging, and global marketing.

Intra-firm trade can cover both final and intermediate goods. Here our focus is intermediate goods. The use of imported intermediate inputs in manufacturing depends on the industrial organization and international trade factors just mentioned, as well as on supply chain management tools that control demand, supply, and quality variability. In fact, global economic integration has allowed MNEs to create fragmented production processes by locating their intermediate production activities in various parts of the world. Together with such fragmented production comes an intensive trade in intermediate inputs for the production of the final manufactured good. [3]

Table 5-1 shows aggregate US data on intermediate-input imports and intra-firm imports, for 1992 and 1997. About three-eighths of all US goods imports are intermediate inputs (the remainder are final goods). About 43% of all US goods imports arrived through intra-firm channels in 1992, rising to 52% in 1997 (the remainder came through arms-length channels).

Table 5-1. Imports into the United States by Trade Categories, As Percent of Total Imports

	1992	1997
Intermediate Inputs/Final Goods		
Percent intermediate inputs	37%	38%
Percent final goods	63%	62%
Intra-Firm/Arm's Length		
Percent Intra-Firm	43%	52%
a) Percent US MNEs	17%	30%
b) Percent Foreign MNEs	26%	22%
Percent Arm's Length	57%	48%
Addendum: Total Imports $ Billions	505	748
Sources: Authors' calculations, see Appendix.		

---

[3] See Arndt and Kierzkowski (2001) for a collection of papers on fragmented production.

The empirical tests in this chapter focus on United States imports, due to the importance of US MNEs in international trade and because the US Bureau of Economic Analysis (BEA) has provided high-quality tabulations of several especially relevant data sets. First, the BEA publishes detailed data on intra-firm trade in goods, involving both US and foreign MNEs and their respective affiliates. These data are based on extensive benchmark surveys taken every five years, as well as smaller annual surveys. The BEA also publishes related data on US foreign direct investment abroad and foreign direct investment in the US. Finally, US imported intermediate inputs can be computed by combining three BEA data sets:

1) An input-output data set, based on the 1992 and 1997 US Census of Manufactures, is applied to determine the total quantity of intermediate inputs by industry.
2) Industry import data are then used to estimate imported intermediate inputs.
3) Import data by industry and country of origin are then used to estimate imported intermediate inputs by country of origin.

The Appendix to this Chapter discusses how we calculate imported inputs by industry and country of origin.

Previous studies have applied only steps 1) and 2) of this methodology. Also, these studies primarily used industry data and focused on labor market impacts or exchange rate exposures[4]. Our analysis, in contrast, focuses on a country cross-section, and applies the data to the interaction of imported inputs and intra-firm trade. Although similar issues arise in both goods and services, in this chapter we confine ourselves to trade in goods alone.

5.2   LITERATURE REVIEW

The issue of intra-firm trade is inextricably linked to multinationals and foreign direct investment (FDI). A large part of the FDI literature deals with its country-wise determinants, such as size, relative endowments, and trade and investment costs (Carr, Markusen and Maskus, 2001), relative rates of return (Chernotsky, 1987), and, in the case of foreign investment in R&D activity, the size of the scientific base (Kuemmerle, 1999). There is also a literature that assesses the impact of FDI on the local, host economy in terms

---

[4] Work on this method was pioneered by Campa and Goldberg (1997), Feenstra and Hanson (1996), and Hummels, Raporport, and Yi, (1997).

of its impact on innovation (Glass and Saggi, 2002) and economic growth (Zhang, 2001, and Nair-Weichert and Weinhold, 2001). For our purposes, the literature that deals with transnational vertical integration and intra-firm trade is of greater relevance. For example, Wilamoski and Tinkler (1999) show that there was a rise of intra-firm exports and imports between the US and Mexico as a result of US FDI in Mexico. Other studies of multinational firms have looked at the motivation behind investment abroad and whether FDI complements or substitutes trade (Konan, 2000, and Roy and Viaene, 1998). Konan's theoretical model, in particular, shows that intra-firm trade in intermediate goods implies that vertical investment complements rather than substitutes for trade.

Another branch of the intra-firm trade literature deals with its determinants. For example, Helpman (1984) develops a model that generates shares of intra-firm trade as a function of relative nation size and variations in relative factor endowments. A large literature also exists on transfer-pricing and taxation issues and their relationship with intra-firm trade (Taylor, 2002), while Madan (2000) shows how different levels of taxation in the host-country give rise to a different mix of intra-firm trade in final and intermediate goods.

Turning to outsourcing, Grossman and Helpman (2002) study the determinants of outsourcing locations in a global economy using a general equilibrium trade model. Costly searches and incomplete contracts are critical in this model. The relative thickness of markets for input suppliers, relative search costs, and the contracting environment impact the extent of global outsourcing. Countries with an active inputs market and reliable contracting environment would be a relatively dependable site for outsourcing. In an empirical study, Andersson and Fredriksson (2000) show that internal imports of intermediate goods by Swedish firms were dependent on their international organization and concentration of production, market-size, and R&D expenditures.

The large size of the literature reviewed in this section confirms the importance attached to the separate topics of intra-firm trade and imports of intermediate inputs. On the other hand, the combination and integration of these two key aspects of globalization appears not to have been studied. This intersection of intra-firm trade and imports of intermediate inputs is thus the focus of our empirical tests, to which we now turn.

5.3   ANALYSIS

To start, it is useful to clarify the relationship between imported intermediate inputs and intra-firm imports. This relationship is illustrated in Figure 5-1. The full 360-degree circle represents the total amount of goods imported by the home country from any given foreign country in any given year. The right hemisphere (quadrants 1 and 2) show intra-firm imports, representing transactions between a MNE and its affiliate, either from a foreign affiliate to a home country MNE, or from a foreign MNE to its home country based affiliate. Intra-firm trade can occur in either intermediate inputs or final goods, represented by quadrants 1 and 2 of the circle respectively. The left hemisphere (quadrants 3 and 4) represents imports from arm's length trading partners, meaning that these imports are not carried out within the same firm. Arm's length trade can also occur in either intermediate inputs or final goods (in quadrants 4 and 3 respectively).

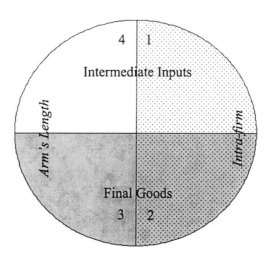

Figure 5-1. Total Imports into Home Country, Classified as Intermediate Inputs or Final Goods and as Intra-Firm or Arm's Length Trade.

The quadrants in Figure 5-1 are of equal size only for graphical convenience. In fact, a primary goal of this study is to determine the average size of these quadrants, and to determine what factors cause the quadrant sizes to vary across countries. As already noted in Table 5-1 for 1997, however, we do have information concerning the size of the two sets of hemispheres in Figure 5-1:

- 38% of all US imports were intermediate goods and 62% were final goods;
- 52% of all US imports were intra-firm imports and 48% were arm's length.

Also, about three-fifths of all intra-firm imports were carried out by US MNEs, the remainder being imports by the US affiliates of foreign multinationals.

The size of the four individual quadrants cannot be derived from aggregate data that only separate intra-firm from arm's length trade and intermediate from final goods trade.[5] In fact, we know of no standard data set that provides separate quadrant sizes. However, the same information that we have illustrated in Figure 5-1 and summarized in Table 5-1 for aggregate imports into the United States are also available on a disaggregated basis by country of origin (i.e. the exporting country). These data have the potential to provide substantially more information about the distribution of import flows across the four quadrants.

As a simple example, assume the split between intra-firm and arm's length trade is 50/50, and the split between intermediate input and final good trade is also 50/50. With no further information, we cannot know the size of each of the four quadrants of Figure 5-1. Now also assume that intermediate input and intra-firm imports always occur together, and that final goods and arm's length imports also always occur together, although the two factors may vary across countries. This pattern is still consistent with an aggregate 50-50 split between intra-firm and arm's length trade and also between intermediate inputs and final goods trade. But the disaggregated patterns provide additional information. In particular, we now know that quadrants 2 and 4 of Figure 5-1 must be empty, and that exactly 50% of the trade would appear in each of quadrants 1 and 3. This example illustrates why data disaggregated by country of origin may provide insights into the aggregate data, which are not available from the aggregate data directly.

It is worthwhile to consider one more example in order to illustrate a slightly more general case. Here we assume that the country data allow us to determine that:

1) 50% of aggregate imports are intermediate inputs and 50% are final goods.
2) All intermediate inputs arrive through intra-firm trade.
3) 50% of final goods arrive as intra-firm trade and 50% arrive as arm's length trades.

---

[5] We need three pieces of independent information, in addition to the total amount of imported goods (the size of the circle), to determine the size of each quadrant. Data that separate (1) intra-firm and arm's length trade and (2) intermediate and final goods trade provide only two pieces of independent information.

In terms of Figure 5-1, this would imply that quadrant 1 contains 50% of all trade, while quadrants 2 and 3 each have 25% of the total trade. Quadrant 4 would contain no trade.

These examples represent, of course, easy cases. In the real world, the best hope is to find that the cross-country correlations for the various import categories are sufficiently informative to allow us to decipher the true structural features at a reasonable level of confidence.

Table 5-2 shows total imports and input imports into the US from the major countries of origin in 1997. The table also provides input imports/total imports ratios and intra-firm imports/total imports ratios by country; for source of all data, see the Appendix. Much greater variation is apparent in the intra-firm import ratios. More than 70% of the US imports from countries such as Japan are carried out through intra-firm trade, while at the other end of the spectrum, imports from Taiwan are primarily of an arm's length nature. The table also reflects the diverse nature of the countries shipping intermediate inputs to the United States, covering developing and developed countries, and European and Asian countries alike: a true testimonial to globalization.

Table 5-3 shows total imports and intermediate input imports for the four industries with the largest amount of total imports among all US 3-digit NAICS industries in 1997. The table also shows the top three countries of origin for intermediate inputs for each industry. Computers and Electronics (NAICS 334) and Transportation Equipment (NAICS 336) are first and second with respect to both total imports and input imports. The input import ratio of 39% for NAICS 334 is somewhat lower than the other industries shown, since many computer and electronic products are fully assembled abroad and then imported as final goods.

Table 5-4 lists the countries of destination of US foreign direct investment in decreasing order of magnitude as of 1997. The preponderance of the "old industrial Europe" is striking, as is the mismatch with the major countries sending imports to the US (see Table 5-2), and with the countries which export high-tech inputs to the US (see Tables 5-3 and 5-5). Further light will be thrown on the origin of imports and their connection to different kinds of firms in the empirical section.

Table 5-5 shows the top ten countries of origin and the corresponding quantities for US high-tech imported inputs for NAICS 334, Computers and Electronic Products Manufacturing. Table 5-6 provides our estimates of total imported inputs and high-tech imported inputs, for both California and the rest of the US. This illustrates California's large share of imported inputs.

Table 5-2. United States Imports by Trade Categories and Major Countries of Origin, 1997.

	Total Imports $ Billions	Input Imports/ Total Imports Ratio	Intra-firm Imports/ Total Imports Ratio
Canada	168	0.40	0.47
Japan	121	0.42	0.71
Mexico	86	0.36	0.35
Germany	43	0.43	0.60
China	63	0.29	0.10
UK	33	0.43	0.47
Taiwan	33	0.43	0.08
France	21	0.46	0.38
Korea	23	0.36	0.22
Italy	19	0.40	0.67
US Total	748	0.38	0.52

Sources: Authors' calculations, see Appendix.

Table 5-3. Imported Inputs into the United States, by Largest Importing Industries, 1997.

NAICS Code	Sector Name	Total Imports $ Billions	Input Imports $ Billions	Input Imports/ Total Imports Ratio	Top 3 Countries of Origin
334	**Computers & Electronics**	**173**	**68**	**0.39**	**Japan, Taiwan, Mexico**
336	Transportation Equipment	149	72	0.48	Canada, Japan, Mexico
333	Machinery ex. Electrical	65	35	0.54	Japan, Canada, Germany
325	Chemicals	51	26	0.51	Canada, Japan, Germany

Sources: Authors' calculations, see Appendix.

Table 5-4. US Foreign Direct Investment, by Selected Countries, $ Billions, 1997.

Country	Investment
United Kingdom	154
Canada	97
Netherlands	69
Germany	41
France	37
Brazil	36
Japan	34
Switzerland	31
Australia	28
Mexico	24

Source: Bureau of Economic Analysis, US Department of Commerce

Table 5-5. US High-Tech Imported Inputs By Top Countries of Origin, $ Billions, 1997.

Country	High-Tech Imported Inputs
Japan	15
Taiwan	8
Mexico	8
Singapore	6
China	5
Korea	4
United Kingdom	4
German	2
Philippines	2
Source: Authors' calculations, see Appendix.	

Table 5-6. Imported Intermediate Inputs, California and Rest of United States, 1997.

	California	Rest of US
High-Tech Input Imports ($ Billions)	17	50
Total Input Imports ($ Billions)	30	295
High-Tech as Percent of Total Input Imports	56.7%	16.9%
Source: Authors' calculations, see Appendix.		

5.4   REGRESSION ESTIMATES

We now turn to a series of regression tests on intra-firm and intermediate input imports into the United States using a cross section of countries of origin. The dependent variable is the log of the dollar amount of intermediate inputs imported into the United States from each of the countries. The empirical descriptions of these and all other data are provided in the Appendix. We estimate multivariate cross-section regressions for the years 1992 and 1997 separately to determine which factors are most highly correlated with the observed cross-country pattern.

Our specification for the independent variables starts with a one-direction version of the gravity model, since we are looking at the imported intermediate inputs from each trading partner to the US and not the bilateral trade among them; see Feenstra, Markusen, and Rose (2001) for a recent survey.[6] We then modify the standard model by including intra-firm and arm's length goods imports by country as additional explanatory variables.[7] We also separate intra-firm trade into imports from the foreign affiliates of US MNEs and imports from foreign MNEs to their US affiliates. In summary, the following are our primary independent variables, computed for each of the countries of origin (in the estimated equations, all variables except the Asian dummy are measured in logs):[8]

- IFUSA   US imports from the foreign affiliates of US-based MNEs.
- IFFOR   US imports from foreign-based MNEs to their US-based affiliates.
- ARML   US imports sent to arm's length recipients (= Total US Imports – IFUSA - IFFOR).
- GDPPC   Gross domestic product per capita of the country of origin.
- POP   Population of the country of origin.
- DIST   Great circle distance between largest city of foreign country and Kansas City, Mo.
- ASIAN   Dummy variable: 1 for Asian countries of Asia-Pacific Economic Cooperation.

The results in Table 5-7 are divided into three sections. Section A has the log of imported intermediate inputs as the dependent variable, and the regression is estimated on a cross-section of forty-eight countries for which data are available for 1992. Equation (1) is a standard gravity model, based on per capita GDP, population (as a measure of size), distance, and a dummy variable for the Asian countries. The adjusted R^2 is over 75%, indicating that an important share of the cross-country distribution of US intermediate inputs can be explained on the basis of gravity variables alone. We also tested a variety of other gravity variables, but none were consistently significant. Our basic results would be unaffected by including any of these variables.

---

[6] The gravity model owes its name and origin to the phenomenon of gravity in physics. Bilateral trade (similar to the force of gravity) is assumed to be proportional to the product of GDPs of the trading countries ("mass" in the physics analogy) and is inversely proportional to the distance between them. As pointed out by Feenstra, Markusen and Rose (2001), the gravity model fits trade data very well and is consistent with many alternative theories of trade. In recent years its usage has been extended to assess the impact of trade and openness on national or regional growth and income (Frankel, Romer, and Cyrus, 1996), and to discover natural trading blocs (Frankel, Stein, and Wei, 1996).

[7] We discuss the gravity model and our modifications in greater detail in Chapter 6.

[8] We also note that the following identity holds among the trade variables (also see Figure 1): Total Imports = Intermediate Input Imports + Final Good Imports = IFUSA + IFFOR + ARML.

Equation (2) in Table 5-7 adds three disaggregated import flows as potential determinants of imported intermediate inputs. Their coefficients measure the elasticity of imported inputs with respect to each of the US import categories. The results indicate that intra-firm trade was primarily related to final good imports as of 1992. We also tested for a direct effect of US foreign direct investment in each country, but it did not provide an independent effect over and above the effect of the related intra-firm trade flows. We also estimated all the equations in Table 5-7 using instrumental variables, but none of the results were changed in any substantive way.[9]

Part B of Table 5-7 repeats the estimation of Part A for 1997 data. The sample size is now only thirty-eight countries, since the 1992 data rely on a special tabulation carried out by the Bureau of Economic Analysis (see Zeile (1997)), and a comparable tabulation is not yet available for the 1997 data. Equation (3) provides estimates based on the gravity specification alone, with results similar to those obtained for the 1992 data. Equation (4) adds the same three import variables used in equation (2). The coefficient estimates for 1997 for these variables indicate a significant increase in the importance of intra-firm trade, related to both US and foreign MNEs, and a corresponding reduction in the importance of arm's length trade, as a determinant of imported intermediate inputs. This is an important result, since it confirms the view that MNEs are increasingly using foreign outsourcing as they decentralize their production processes.[10]

Part C of Table 5-7 repeats Parts A and B (with 1997 data), but the dependent variable is now the log of imports of only *high-tech* intermediate inputs, defined here as NAICS code 334.[11] Equation (5) begins with the gravity model variables, with two notable differences from the results in equations (1) and (3). First, the distance variable is now much less important. This is understandable since high-tech imports are commonly referred to as "weightless" in terms of their value-to-weight ratio implying that they are much less sensitive to transportation costs. Second, the Asian country dummy is much more important than it is in the earlier equations. This too makes sense, since there is other evidence that the Asian countries are of increasing importance as sources of intermediate inputs for US high-tech industries.

---

[9] The two instruments were: (1) an index of competitiveness that took into account the investment climate, and availability of skilled labor among other measures, and (2) the share of high-tech exports in total exports.

[10] It is useful at this point to note again that our import data correspond only to goods imports.

[11] Specifically, NAICS 334 is defined as Computers and Electronic Product Manufacturing, and includes semiconductors, scientific instruments and telecommunications equipment.

Table 5-7. Regression Results

Eq #	Cons-tant	IFUSA	IFFOR	ARML	GDPPC	POP	DIST	ASIAN	Adj R²
Part A: Imported Intermediate Inputs is Dependent Variable--1992 Data									
1)	-5.31* (2.15)				.92* (11.94)	.77* (8.99)	-1.00* (3.97)	1.90* (7.27)	0.78
2)	-5.31* (2.56)	.01 (.12)	.06 (.74)	.80* (5.34)	.49* (2.72)	.18 (1.65)	-.19 (.77)	.47 (1.18)	0.89
Part B: Imported Intermediate Inputs is Dependent Variable--1997 Data									
3)	-7.96* (2.51)				.93* (9.15)	.87* (8.38)	-.89* (3.8)	1.65* (4.88)	0.76
4)	-2.93* (1.98)	.17* (2.98)	.24* (4.53)	.36* (6.17)	.24* (2.97)	.22* (2.89)	-.13 (.90)	.46* (2.63)	0.95
Part C: High-Tech (NAICS 334) Imported Intermediate Inputs is Dependent Variable--1997 Data									
5)	-27.9* (3.36)				1.54* (5.96)	1.34* (5.46)	-0.55 (.98)	3.39* (4.15)	0.58
6)	-23.2* (2.33)	.77* (2.50)	.38 (1.11)	.06 (.26)	.50 (1.12)	.39 (.93)	.92 (1.74)	1.54 (1.87)	0.73

Notes:
Ordinary Least Squares with White heteroskedasticity adjustment.
Absolute values of t-statistics shown in parentheses; * significant at 5% level.
All regressions are estimated on a cross-section of countries, 48 countries in 1992,
38 countries in 1997.
All data are in logs except for the Asian dummy.
See Appendix for detailed description of data series.

Equation (6) of Table 5-7 adds the imports variables to the basic gravity model for high-tech imported inputs. Compared with equations (2) and (4), equation (6) indicates that US imports of high-tech intermediate inputs depend primarily on intra-firm trade (not arm's length transactions), and especially on imports by US MNEs. Indeed, imports by US MNEs are now the predominant source of imported high-tech intermediate inputs into the US. This result provides empirical verification of the view that foreign outsourcing has become especially important for US MNEs in high-tech industries.

5.5   CONCLUSIONS

High and growing levels of intra-firm trade and intermediate input trade play key roles in the new era of globalization. Although they have been intensively studied individually, little attention has been paid to their interaction, i.e. intra-firm trade in intermediate inputs. A major problem has been the lack of data that measure the amount of intra-firm trade that involves intermediate inputs, or vice versa. This chapter offers two primary innovations. First, we have developed a data set of imported intermediate inputs by both industry and country of origin. Second, we have used estimates from a regression model to determine the importance of intra-firm imports as a determinant of trade in intermediate inputs.

Our key results, from Table 5-7, are:

1) Intra-firm imports were a relatively unimportant source of intermediate imports as of 1992. Most US intermediate goods imports at that time were the result of arm's length trades.
2) By 1997, intra-firm trade, by both US and foreign MNEs had become very important as a source of imported intermediate inputs. However, arm's length trade also remained a significant determinant of US intermediate input imports.
3) Standard gravity model variables were found to be important determinants of US imports of intermediate inputs, in addition to the key role of intra-firm trade variables.
4) Estimates were also derived for *high-tech* intermediate input imports, defined as NAICS code 334, which represents computers and electronic products. These additional results were:
   a) Transportation costs, measured by distance, were not a major hindrance to high-tech intermediate imports, consistent with the high-value, low-weight, character of these goods.
   b) Intra-firm trade (not arm's length transactions), especially imports by US MNEs, is a key determinant of high-tech intermediate input imports, consistent with the view that foreign outsourcing has become especially important for US MNEs in high-tech industries. In particular, US MNEs were responsible for more than two-thirds of all imports of high-tech intermediate inputs into the US.

Point (4) has special significance for a state such as California with an intensive high-tech economy. Not only is foreign outsourcing important for the state's high-tech firms, but these firms are likely to be central to the importing of high-tech products into the United States.

5.6 APPENDIX: DATA DESCRIPTION

5.6.1 Computation of Imported Inputs for the United States, by Country of Origin

To calculate imported intermediate inputs by sector and by country of origin, we applied the following formulas to each 6-digit input sector in US manufacturing (all amounts in $ Billions):

$$(1) \quad II_i = I_i \left(\frac{M_i}{(P_i - X_i + M_i)} \right)$$

where

$I_i =$ amount of sector i goods used as inputs in all of US manufacturing (from US Census of Manufacturing Input/Output data for 1992 and 1997 respectively).

$II_i =$ imported inputs of sector i goods;

$M_i =$ total imports of sector i goods;

$P_i =$ US production of sector i goods;

$X_i =$ US exports of sector i goods.

The basic assumption here is that, for any input sector, the percentage that imports of intermediate input represent of total intermediate inputs is the same as the percentage that imports represent of all net sources of that commodity ($= P_i - X_i + M_i$).

$$(2) \quad II_{ic} = M_{ic} II_i.$$

where

$II_{ic} =$ Imported intermediate inputs of sector i from country c.

$M_{ic} =$ Sector i imports from country c as a proportion of US total sector i imports.

The basic assumption here is that country c's share of imported intermediate imports of sector i goods equals that country's share of all imports of sector i goods.

5.6.2 Computation of Imported Intermediate Inputs for California Manufacturing

Imported intermediate inputs for California are computed from the US intermediate input numbers, using the following equations:

$$SI_{Ca} = I_{Ca} / I_{US}$$

where,

SI_{Ca} = Share of US inputs used in California manufacturing;
I_{Ca} = Inputs used in California manufacturing;
I_{Us} = Inputs used in US Manufacturing.

$$CAII = SI_{Ca} * II$$

where,

CAII = Imported inputs used in California Manufacturing,
II = Imported inputs used in all of US manufacturing from equation (1) in Appendix, summed across all sectors.

5.6.3 Data Sources

Trade data by countries and industries
All import data by countries and industries are from US International Trade Commission's Trade DataWeb web site: http://dataweb.usitc.gov.

Intra-Firm Trade Imports by Country of Origin
Data for 1992 are from Zeile (1997);
Data for 1997 are from Mataloni (1999) for US MNEs and from Zeile (1999).

Gravity Model Variables
The distance data have been calculated using Encarta. Gross Domestic Product (GDP), Gross Domestic Product per Capita (GDPPC) and Population (POP) are from the World Bank database:
http://devdata.worldbank.org/data-query.

Chapter 6

International Networks and High-Tech Exports

Although the geography of international trade and its spatial distribution have traditionally been analyzed at the level of the nation-state, for a large country, such as the US, this aggregate level of analysis may sometimes obscure some underlying regional economic features and attributes. The latter have acquired increasing significance, since growing global economic integration has given an added impetus to forces of agglomeration and specialization within nations. For example, export data available for constituent states of the US, allows us to study variations in trading patterns among different states and regions of the union, as well as the differences in their economic determinants. The study of international trade of sub-national regions offers us the possibility of gaining insights into the building blocks of macro trading affinities of nations of which they are a part. The trading affinities of regional economies would also be helpful in assessing their vulnerability to foreign shocks and in determining changes in competitiveness relative to the national and other sub-national economies.

Our objective in this chapter is to study the determinants of high-tech and general goods exports from two regional economies comprising the national economy of the United States – the economy of the state of California (CA) and the economy of the rest of the US (RUS), and the differences between them. Part of the motivation for the present book and this chapter arose out of the increasing emphasis placed by regional and state authorities on stimulating exports (see Chapter 7 in this book). The number of International Trade and Investment offices of the state of California, for example, as well as of similar export promotion offices of various other states of the union had been increasing regularly until recently. States frequently schedule trade missions led by senior officials. Indeed, there has been a trickle-down effect of this policy initiative on the part of state officials, with city mayors now scouting for world markets for their metropolitan exports. The primary factor behind this emergent enthusiasm for trade has been the realization, particularly in the more open coastal states, that exports are increasingly a very sizable part of the gross regional product.

This chapter has also been motivated by the size of the economy of California, its openness to trade, the large presence of a globalized high-tech sector, as well as its location on the western seaboard and its linkages with the global economy. For comparative purposes we also analyze in a similar manner, the export patterns of the rest of the US (RUS). Studying the differences between these two regions in their trading affinities and trading patterns can contribute to an understanding of the dynamic between national and sub-national economies in the age of increasing globalization.

In this chapter we study both the exports and other global linkages of these large regional economies. The coastal states, California in particular, are more open not just to trade and investment flows, but also to the impact of international social (immigrant) and business networks. A significant proportion of the flow of immigrants arrives in the coastal states of the US; but more importantly, a sizable number of them choose to settle down there. The ease of international travel and the entrepôt nature of the coastal ports also leads to a high concentration of transnational business linkages based along the coasts, in the form of foreign direct investment, as well as various other forms of business-to-business contacts. We aim to connect all these disparate strands into a story of global linkages and their impact on trade in California and the rest of the US.

Our methodological tool is the gravity model, a widely used method of studying international and interregional trade relationships, first introduced in Chapter 5. Instead of applying a traditional gravity model to the US as a whole, we apply a one-sided version of the model to total goods exports of California, as well as its computer-related exports, and to those of the rest of the United States. We also modify the traditional gravity model to check for the influence of other quasi-economic trading stimuli such as transnational social ties and networks and international business networks in determining the direction and volume of exports.

By transnational social ties and networks we mean the contacts that bind foreign-born immigrants to their countries of birth; international business networks are understood as intra-firm commercial transactions connecting affiliates, subsidiaries, offices and headquarters across national borders. The exact formulations are explained later. In the context of a gravity model, the choice of sub-national regions instead of the nation as an object of study helps us arrive at a more accurate measure of distance with the regions' trading partners. It also allows us to test for differential trading affinities of two constituent regions of a large nation, whose geographical centers are far removed from each other, and which are geographically closer to different parts of the world: the Asia-Pacific region in the case of California (CA), and Europe/South America in the case of the rest of the US (RUS).

The questions we address in this chapter are:

1) What are the country-wise determinants of high-tech goods exports from CA and RUS, where high-tech is defined as computers, peripherals and electronic products?
2) How do business and socio-cultural networks influence the level of exports from CA and RUS to different countries?
3) Does either region have an "affinity" to trade with particular parts of the world?

The study of trading patterns of sub-national regions, while accounting for the ties and networks that bind local and regional economies to the larger global economy, provides an integrated view of the regional economy as a part and parcel of the global economy.

6.1   LITERATURE REVIEW, GRAVITY MODEL, AND NETWORKS

Rapid advances in information and communications technologies and intensifying global economic integration are rendering the nation-state somewhat less meaningful as an economic unit. Paul Krugman (1991), among other economists, has been advocating an integrated look at regional economics, macroeconomics, and international economics, under the aegis of economic geography.

The economic strands connecting regional economies to the global economy have usually been looked at through the prism of regional competitiveness, development of local industry, and impact on local labor markets. Shift-share techniques have been usefully employed to assess trade related job changes at the regional level, as well as for evaluating regional advantage (Naponen, Graham, and Markusen, 1993, Sihag and McDonough, 1989). However, the actual global geographic distribution of exports of regional economies has not been widely studied.

6.1.1   Immigrant Ties

Although the impact of foreign-born immigrants on issues relating to wages, employment, welfare and business formation has been widely studied

(Simon, 1992), their role in international trade is only now receiving long-overdue attention. Social scientists have thrown considerable light on the creation and maintenance of, and role played by transnational social networks formed due to ties with the country of origin of foreign-born immigrants in the US. Indeed, the determinants of immigrant concentration and the micro-structure of migration networks have been studied extensively (Castles and Miller, 1993; Massey and España, 1987; Belanger and Rogers, 1992; and Bartel 1989). Several studies have also analysed the role played by these networks in sustaining business and economic ties with the country of origin (Borjas, 1994; Matthei, 1996; Ho, 1993; and Saxenian, 1999).

The relationship between immigrant populations and trade is an emerging area of research in economic and sociological literature. A number of studies have shown that the size of the foreign born populations contribute significantly to trade between the country of origin and the adopted country (Gould, 1994; Bardhan and Howe, 1998; Head and Ries, 1998). Rauch (2001) reviews this body of literature and suggests that there exists evidence pointing to the role played by transnational business and social networks in promoting international trade due to mitigation of contract enforcement problems and enhancing information flows. The increasing share of differentiated products in international trade further tended to bolster the role of these networks. Gould's 1994 study of US trade with forty-seven partner countries showed that immigrants facilitated in expanding trade to their countries of origin with exports being more strongly influenced than imports. A similar study of Canadian trade with partner countries reaffirmed the positive influence of immigrants on trade (Head and Reis, 1998).

Dunlevy and Hutchinson (1999) looked at the historical impact of immigrants on American imports for the period between 1870 and 1910. This study, which examines the contribution of migrant stock to imports from their countries of origin, disaggregated both by exporting regions and by commodity type, corroborates the trade enhancing effects of immigrants in the more recent cases reported earlier. The results show considerable variation in the influence of migrants by their region of origin on US imports, and note that the impact is significant only for finished and differentiated goods such as manufactures and foodstuff.

The evidence provided in this literature suggests that there exists a positive relationship between migrant flows and trade flows. However, more disaggregated analyses have shown that this relationship is conditioned by a number of contextual factors. Given that studies have only been available for two countries (US and Canada) and that most of these studies were conducted at the national level, generalization of these results is limited. Questions remain regarding the robustness of these results. It is possible, for example, that some key geographical region of the US accounted for the results, or that the variation could have been explained by the type of immigrants and by the type of commodities traded.[1]

6.1.2 Business to Business Ties

A different branch of international trade literature focuses on another set of networks that may impact trade flows. These are business networks established as a result of transnational investment and business-to-business contacts. As mentioned in Chapter 5, studies of multinational firms have looked at the motivation behind investment abroad and whether foreign direct investment (FDI) complements or substitutes trade between the country of origin of investment and the host country (Konan, 2000; Roy and Viaene, 1998). Wilamoski and Tinkler (1999) show that there is a rise of intra-firm exports and imports between US and Mexico as a result of US FDI in Mexico. Similarly, other studies have found that intra-firm trade in intermediates implies that vertical investment complements rather than substitutes for trade (Konan, 2000). Clausing (2000) finds evidence that supports the conclusion that multinational activity and trade are complementary activities.

However, most of the empirical literature on this topic looks at the issue of complementarity or substitutability between *local sales* in a foreign country of foreign affiliates of US multinationals, and *exports* from the US to that foreign country. The idea in those papers is to examine whether those *local sales* by multinational affiliates displace the *exports* that were previously being shipped from the home country. The literature largely does not examine a related issue—whether exports by multinationals from the home country augment the overall total exports from the home country, i.e. whether there is a positive spillover effect of intra-firm trade on arm's length trade.

Studies of intra-firm activities have also focused on transfer pricing issues and on the connection between foreign direct investment and trade. The magnitude of intra-firm trade by country has become increasingly evident with the publication of survey articles and data by the US Department of Commerce in 1997 (Zeile, 1997). The large proportion of intra-firm trade signifies a strengthening and deepening of cross-country business ties. Inclusion of a variable as a proxy for these networks in a gravity model can help us ferret out the spillover effects that within-firm trade can have on arms-length transactions, and hence on total trade. We test for complementarity between intra-firm trade and exports for both California and RUS. Since we use the intra-firm export ratio as a variable, the complementarity in question is actually between the proportion of intra-firm exports in total exports from the two sub-regions of the US, on the one hand, and the total exports of these

---

[1] See Rauch (1999).

two regions, on the other, i.e. we specifically test for a spillover effect from intra-firm exports to arm's length, and hence overall exports.

6.2   THEORETICAL CONSIDERATIONS

International economists have derived the gravity equation, which predicts that bilateral trade between two countries would be proportional to the product of their respective outputs and declining in distance between them, from basic principles of international economics. In one of the better-known papers in the field, Deardorff (1995) shows how the basic Heckscher-Ohlin model of international trade can lead to a gravity specification for bilateral trade. When countries produce distinct goods with complete specialization or for a whole range of other assumptions and preferences, trade takes place according to the standard gravity model and declines with distance, and the departures therefrom are due to relative transportation costs. Harrigan (2001) reviews the theoretical and empirical literature on gravity models, and similarly stresses the role played by relative as well as absolute transportation and other trade and transactions costs. In the empirical literature, the extensions to the standard version of the model have generally come in the form of variables that would impact trade, transportation and transactions costs. Regional dummy variables have been used to capture both the relative transportation costs of trading with a particular world region, as well as to highlight trading affinities. The rationale for using foreign-born immigrants in the gravity model is precisely because their presence alleviates the costs of doing business between their country of origin and the host country.

Immigrants impact trade both because of their kinship ties and their implicit knowledge about the customs, language, and social norms of their country of origin. Immigrants can substantially reduce the hidden costs of foreign trade by identifying potential customers and business partners and by deftly navigating the plethora of local customs, laws and business practices in their previous home country. In fact, the social and economic conditions, which give rise to issues of trust at the national level, are more pronounced in the sphere of international trade. If agents cannot indulge in complete contracting even in the domestic market with far better information, the legal machinery and reputational threats at their disposal, one can conjecture that international markets are fraught with serious risk in terms of incomplete contracts, economic volatility and high transactions costs. In other words, "economizing on transactions costs", a concept that lies at the core of the theories of Williamson (1998), seems to be one appropriate theoretical framework for understanding the relationship between immigrants and trade. Also related to these issues is the proposition that immigrant populations capitalize on the ethnic capital embodied in their communities (Borjas,

1995), and one of the manifestations of this "capitalization" is expressed in ties and trade flows with country of origin. Insofar as trade is information dependent, in the sense of knowledge about foreign markets, regulations, and contacts, then the pool of immigrants in the US constitutes a large information bank that can be drawn upon to tap foreign markets.

In our case, we bring two different strands of literature together, by including a variable that captures transnational business ties as well. In so far as business-to-business ties enhance business information flows across countries and reduce transactions costs, one would expect them to impact the prices faced by the importers, and consequently the amount of trade between any bilateral set of countries. Unlike the literature cited earlier, the approach here is somewhat different. In this paper, a variable that proxies for transnational business-to-business ties is used to test if these ties have a positive and significant impact on overall exports from the two regions. The variable used is the proportion of exports to different countries that is carried out by multinationals, or the ratio of intra-firm exports to total exports.

We bring together these two independent strands of literature on the impact of immigrants and business networks on trade flows in the context of exports of sub-national regions of the US. Because of the variation in the type and composition of foreign-born immigrants settled in the two regions, as well as the differences in the industrial paths taken by the geographically and historically distinct sub-national economies, it is possible to simultaneously account for both of these effects and assess the differential impact these kinds of global linkages have on their international trade flows.

We estimate the following equation for California and RUS exports:

$$\text{Log (Exports}_i) = \text{constant} + \alpha\text{Log (GDP}_i) + \beta\text{Log (GDP per Capita}_i) + \gamma\text{Log(Distance}_i) + \eta(\text{Openness}_i) + \lambda\text{Log (Foreign Born}_i) + \theta(\text{Intrafirm Export Ratio}_i) + \text{RegionDummy}_i$$

Our version of the gravity model is built on exports alone, since import data are not available for sub-national regions of the US, and is a modified version of the standard model since we are not looking at bilateral, paired trade. Pooled OLS regressions with fixed effects were carried out for the cross-sectional panel of importing countries. Pooled, time-dummy fixed effects specification was chosen to neutralize year specific effects for the three years of data 1998-2000. The dependent variable is the log of goods exports.

Some further clarifications are in order. Four of the independent variables are in log form, which helps in interpreting the estimated coefficients as elasticities. Since the intra-firm export variable is already expressed in percentage terms, we follow general convention in many log-linear models and retain it in its untransformed state. In order to isolate the effect of networks, we need control variables apart from those already included, namely GDP and distance. Frankel, Stein and Wei (1995) have noted that as countries "....become more developed they tend to specialize more and to trade more...". We control for this level of development by including a GDP per capita variable, along with a measure of the trade openness of the countries in our dataset.

In addition, like in many trade-related papers using the gravity model, we include regional dummies. The latter are in binary form and take the value one, in the case of the Asia-Pacific region dummy, whenever the country in question belongs to one of the following: Japan, China, Hong Kong, Korea, Taiwan, Singapore, Malaysia, Indonesia, Philippines, Thailand, Vietnam, Australia or New Zealand, and zero otherwise. Although the Asia-Pacific trade block does not formally exist as yet, papers by Frankel, Stein and Wei (1995) and others have pointed out that, by nature of its intra-regional trade and ties, it does so in a de facto sense. In a similar sense, the European union formalized the already existing intra-regional ties and gave them a further boost. The Europe dummy takes the value one when the country belongs to the European Union, and zero otherwise. These dummies are of course different from the time dummies for 1998, 1999 and 2000 used in a fixed effects sense, where only the constant in the regression, or the intercept, is affected.

6.3   THE DATA

6.3.1   Exports

Exports for California have been calculated from state-level export data available from the International Trade Administration of the US Department of Commerce, which has an arrangement with the Massachusetts Institute for Social and Economic Research (MISER). MISER has been producing the state export database since 1987, under an agreement with the US Census Bureau's Foreign Trade Division.[2] Export data for California and RUS, for the years 1998-2000, are compiled both by country of destination as well as by goods category. The total number of countries in our sample is fifty-eight, accounting for more than 95% of total exports worldwide by these regions. Since no import data are available for states, only export equations are esti-

---

[2] Known as the Origin of Movement series, it is arrived at with a procedure involving shippers' export declaration documents at Customs points and ports of exit. They also use an imputational algorithm to allocate missing data to states. See http://www.ita.doc.gov/td/industry/otea/state/ .

mated. Table 6-1 lists the top export markets for California, for all goods as well as for high-tech goods exports.

Table 6-1. California Top Export Markets.

	California Total Goods Exports ($ Thousands)		
	1998	1999	2000
Canada	12,644,137	13,245,900	15,161,966
Mexico	10,798,216	12,230,871	14,404,472
Taiwan	5,577,928	6,528,739	9,360,348
South Korea	4,005,102	5,890,721	9,237,690
United Kingdom	5,361,838	5,195,492	6,473,729
Germany	4,369,206	4,232,625	6,316,147
Netherlands	3,569,470	4,165,084	5,557,306
Singapore	4,368,098	4,515,644	5,302,287
Malaysia	2,590,602	2,509,917	3,854,018
	California Computer and Electronic Exports ($ Thousands)		
	1998	1999	2000
Canada	5,870,075	6,374,340	7,589,547
Japan	5,214,843	5,059,666	6,805,181
Mexico	4,110,934	4,882,780	5,880,278
South Korea	2,031,030	3,394,677	5,262,588
Taiwan	2,493,677	2,820,113	4,135,843
Netherlands	2,210,392	2,818,231	4,040,513
United Kingdom	2,791,790	2,725,910	3,615,013
Germany	2,018,934	1,925,691	3,414,018
Malaysia	1,927,477	1,948,874	3,072,590
Singapore	3,084,648	2,855,903	3,061,744

Source: International Trade Administration, US Dept. of Commerce; Massachusetts Institute for Social and Economic Research. See http://ese.export.gov/ITA2002/Intro_NEW.htm .

6.3.2   Social Networks

The rationale for including immigrants as an explanatory variable for trade equations in a standard gravity model arises primarily from lowered transactions costs due to ties and contacts in the home country. Insofar as trade is information dependent, in the sense of knowledge about foreign markets, regulations, and contacts, then the pool of immigrants in the US constitutes a large information bank that can be drawn upon to tap foreign markets. Since the contacts and information networks of immigrants in their home country would presumably wither away after a few generations, we incorporate only zero-generation or foreign-born immigrants with a "fresh memory". We use data from Table 144 of the "Social and Economic Characteristics" series of the 1990 US Census that publishes number of foreign-born residents, i.e. the stock of foreign-born immigrants in the US by each state of the Union and by country of birth. In the case of a number of newly independent countries, such as the successor states to Yugoslavia and the Soviet Union, we have allocated the numbers from the parent state to the successor states in proportion to the population. (See Table 6-2 for major countries of origin of foreign born residents in CA and RUS).

Table 6-2. Foreign Born US Residents, in California and Rest of US.

Foreign-Born Residents in Rest of US			Foreign-Born Residents in California		
Country	Number	% of Total*	Country	Number	% of Total*
Mexico	1,823,866	13.41%	Mexico	2,474,148	41.24%
Germany	607,504	4.47%	Philippines	481,837	8.03%
Canada	594,443	4.37%	China	211,263	3.52%
Italy	532,295	3.91%	South Korea	200,194	3.34%
Philippines	430,837	3.17%	Canada	150,387	2.51%
United Kingdom	369,363	2.72%	Guatemala	135,675	2.26%
South Korea	368,203	2.71%	United Kingdom	135,391	1.74%
India	366,617	2.70%	Germany	104,425	1.74%
Dominican-Rep.	344,323	2.53%	Japan	97,554	1.63%
Jamaica	322,936	2.37%	India	83,789	1.40%

Source: US Bureau of Census. * Column 3: Percent of total foreign-born residents in Rest of US; Column 6: Percent of total foreign-born residents in California.

6.3.3   Transnational Business Networks

As a proxy for the extent of business ties connecting the two sub-regions with our sample of countries, we use intra-firm export ratios calculated by the Bureau of Economic Analysis of the Department of Commerce in the February 1997 edition of the *Survey of Current Business*. The intra-firm export ratios by country is the proportion of US exports to that country transacted by US parent firms to their affiliates in that country and by US based foreign affiliates to their parent firms outside the US. The ratios range widely by country, from highs of 74% for Switzerland and 70% for Japan, to less than 3% for Greece, Turkey and Poland.

6.3.4   Distance

We use the traditional great circle distance between the west-central region of California for the state of California, and Kansas City, Missouri for RUS, and usually the largest city of the other country. In countries that cover large geographic areas, such as Russia, India, Australia, and China, the geographic spread of economic activity was taken into account.[3]

6.3.5   GDP, GDP per capita, Openness

Data on the GDP and GDP per capita were obtained from the online database of the World Bank (see http://devdata.worldbank.org/data-query). Our measure of openness to trade is from the Penn World Tables, accessed through the National Bureau of Economic Research (see www.nber.org).

6.4   RESULTS

Table 6-3 shows the results of regressions where the dependent variables are high-tech exports from California (CA) and the rest of the US (RUS). The first model includes the basic gravity variables. The intriguing result here is that for CA, the transportation cost proxy—distance—is not significant when it comes to the exports of high-tech goods. Indeed, the coefficient on distance is insignificant in all specifications for these "lightweight" exports of CA.

In the augmented model, foreign-born residents have a positive effect in both regions: a 1% increase in foreign born immigrants leads to an increase of about 0.35% in exports of both regions. Our initial hypothesis and rationale for including the foreign born variable in the regression held that foreign born immigrants embody knowledge of foreign markets and that California and the US would likely gain from this extra "competitive" edge in penetrating those markets. This result suggests that having a foreign born population resident in the state does lead to gains in the sense of increased exports. The coefficient loses its significance in the California model however, once the regional dummies are included. This does not mean that the immigrant networks do not matter in this specification. Since most of the foreign-born in California are from the Asia-Pacific region, the latter dummy tends to capture the positive effect and nullifies the significance of the coefficient on the foreign-born variable. In any case, it is clear that the high-tech exports of both regions have an affinity for the Asia-Pacific region, as well as NAFTA.

---

[3] We used Microsoft's Encarta software for calculating the distances. The goal is to measure distances from the California of US geographic center of economic activity to the corresponding center in the other country.

Table 6-3. Regression Results: Log of Computer and Electronic Product Exports is Dependent Variable

	California			Rest of US		
Log GDP	0.94*	0.71*	0.81*	0.71*	0.58*	0.65*
	(17.42)	(10.28)	(12.26)	(13.02)	(9.77)	(11.35)
Log GDP per Capita	0.09*	0.12*	0.07*	0.05	0.06*	0.04
	(2.63)	(3.69)	(2.72)	(1.77)	(2.00)	(1.68)
Log of Distance	-0.24	0.40	0.22	-0.35*	-0.05	-0.38*
	(-0.97)	(1.53)	(0.83)	(-2.61)	(-0.37)	(-2.32)
Openness Index	0.47*	0.44*	0.48*	0.35*	0.35*	0.38*
	(7.39)	(7.27)	(8.89)	(5.34)	(5.78)	(6.85)
Log Foreign Born		0.35*	0.05		0.36*	0.18*
		(4.88)	(0.79)		(4.61)	(2.66)
Asian dummy			1.30*			1.06*
			(5.34)			(4.66)
European dummy			-0.47*			-0.64*
			(-2.27)			(-3.25)
NAFTA dummy			2.02*			1.08*
			(4.26)			(2.31)
Adjusted R-squared	0.78	0.78	0.85	0.66	0.68	0.78

Notes: Pooled least squares with fixed effects for years; coefficients on year constants not shown; * means significant at 5% level; t-stats in parentheses; all variables in logs, except openness and dummies.

Previous gravity models have found little evidence for a NAFTA effect for the US as a whole, despite the existence of a Canada-US Free Trade Pact of 1989 and NAFTA itself, which became operational in 1994 (Krueger (2000)). The strong impact of the Asian dummy, which is effectively the Asia-Pacific Economic Cooperation (APEC) zone, less Canada and Mexico, is puzzling because a formal agreement is still in the works and is expected to be concluded only by 2020. It is possible this affinity is due to the extent of trade in high-tech intermediate inputs that is carried out between the US and these regions. Europe apparently plays a less than "normal" role in this particular industrial sector. High-tech exports from both CA as well as RUS are significantly less to this region after taking into account the standard variables.

Table 6-4 shows the results of pooled OLS regressions for total goods exports for the two regions. The first three model specifications and the independent variables are the same as in the previous table. Generally speaking, transportation costs seem to matter more for general goods than for high-tech products in the case of California. The picture of regional affinities remains the same as in Table 6-3, whereas the impact of the transnational social networks in the form of foreign born residents is more mixed here than in Table 6-3.

Table 6-4. Regression Results: Log of Total Goods Exports is Dependent Variable

	California				Rest of United States			
Log of GDP	0.85*	0.64*	0.76*	0.65*	0.68*	0.60*	0.67*	0.54*
	(18.57)	(11.19)	(14.61)	(9.36)	(16.53)	(13.14)	(15.59)	(9.95)
Log of GDP per Capita	0.75*	0.10*	0.06*	0.06*	0.03	0.03	0.02	0.02
	(2.68)	(3.94)	(2.81)	(2.98)	(1.335)	(1.49)	(1.31)	(1.45)
Log of Distance	-0.72*	-0.16	-0.52*	-0.07	-0.60*	-0.43*	-0.62*	-0.36*
	(-3.48)	(-0.76)	(-2.44)	(-0.30)	(-5.95)	(-3.92)	(-5.14)	(-2.90)
Openness Index	0.33*	0.30*	0.34*	0.32*	0.18*	0.18*	0.22*	0.17*
	(6.18)	(5.95)	(7.74)	(7.05)	(3.61)	(3.82)	(5.32)	(4.15)
Log of Foreign Born		0.33*	0.06	0.10		0.21*	0.07	0.10*
		(5.60)	(1.11)	(1.66)		(3.47)	(1.39)	(2.01)
Asian dummy			1.29*	1.05*			0.67*	0.42*
			(6.69)	(5.54)			(3.93)	(2.67)
European dummy			-0.47*	-0.53*			-0.66*	-0.64*
			(-2.87)	(-3.26)			(-4.47)	(-4.52)
NAFTA dummy			1.34*	1.53*			1.00*	1.29*
			(3.55)	(4.09)			(2.88)	(3.98)
Intra-Firm Trade ratio				0.90**				1.00*
				(1.97)				(2.62)
Adjusted R-squared	0.78	0.80	0.87	0.86	0.71	0.71	0.82	0.81

Comparing the results for CA with those for RUS, one can notice another consistent theme, i.e. GDP per capita of foreign countries is by and large not significant for RUS, but positive and significant for CA in each and every specification. The coefficient on the distance variable behaves in a somewhat idiosyncratic way; for example, for CA the coefficient loses its significance once the foreign-born variable is introduced, suggesting that distance, and consequently transportation costs, do not have their traditional negative drag on exports once the social networks are taken into account. The behavior of the coefficient on the distance variable cannot be explained by any significant degree of correlation between number of foreign-born immigrants and distance, which would explain why the latter drops out after inclusion of the former (see Table 6-5 for sample correlations).

Table 6-5. Sample Correlations

	GDP	California Exports	GDP per Capita	Foreign Born	Distance
GDP	1				
California Exports	0.67	1			
GDP per Capita	0.19	0.24	1		
Foreign Born	0.09	0.48	-.09	1	
Distance	-0.03	-0.018	0.063	-0.29	1

Practitioners of gravity models have acknowledged that the principal natural barriers to trade may be cultural, or perhaps logistical, in origin and it was thought that variables for distance or proximity could adequately reflect such intangibles (Frankel, Stein, and Wei, 1996). One could conjecture that our results have to do both with the fact that contacts and ties to the home country are themselves independent of distance, and that frequent travel home serves more than one purpose, thus spreading out the fixed costs. Excluding the data for Mexico, the major contributor to California immigration numbers, does not significantly change the results, either for these regressions, or for those in Table 6-3. But since the coefficients on distance and foreign-born variables are not particularly robust to different specifications, one would not like to make too much of it.

The fourth specification includes our measure of business ties across borders – the intra-firm export ratio. It is positive and significant for both CA and RUS, suggesting that there are strong complementarities and spillover effects from intra-firm trade for exports overall. The greater the proportion of total trade that is conducted within multinational firms, the greater the overall exports to the country where the affiliate/headquarters are located, i.e. the amount of arm's length exports seems to be positively impacted by intra-firm exports from the two regions. The theory of intra-industry trade suggests that like trades with like, or that most of the trade of a developed economy is with other developed economies because of the similarity in their industrial structures, leading to intense trade in differentiated goods and intermediate inputs for complex production processes. Most intra-firm trade is often also intra-industry trade, except in the case of firms-conglomerates or international trading companies that deal in a whole range of goods. But although we have controlled for the standard of living with the per capita variable, the coefficient on the intra-firm export ratio variable is robustly positive and significant. For an explanation, one has to therefore look deeper into the kind of goods being traded.

As mentioned in Chapter 5, international trade in intermediate inputs explains a large proportion of within-firm trade. Global economic integration has led to a proliferation of plants manufacturing intermediate inputs in an increasingly diverse set of countries. The traditional cross-border intra-firm trade involving only the rich countries, the UK, Netherlands, US and Japan is now being supplemented by more dispersed intra-firm trade. (See Arndt and Kierzkowski (2001), as well as Chapters 3 and 5 in this book). The story involving trade in intermediate inputs may explain why the Asian effect remains significant in the fourth set of regressions, albeit with a diminished magnitude, suggesting that it cannot be intra-firm trade with this region that accounts for this trading affinity, but possibly some kind of "complementarity" of industrial structures, expressed in cross-Pacific supply chains and production networks.

6.5 CONCLUDING REMARKS

The gravity model seems to fit sub-national export data quite well. Indeed, *a priori* there is no reason to believe otherwise, although one can conceivably think of small enough sub regions for which the transportation costs might vary more idiosyncratically due to specialization and supply relations.

The two large sub-national economies, California and the rest of the US, appear to have somewhat similar trading affinities, although California seems to possess a greater propensity for trade with the Asia-Pacific region. It would appear that variables used as proxies for social and business contacts are just as important as the minimalist geographical attribute of distance, at least for California, and transportation costs seem to matter less, when some other "gravitational" forces are taken into account. In all likelihood, this is due to a combination of many factors, such as trade in intermediate inputs, particularly of an intra-firm nature, linking the US West Coast with the Asia-Pacific economies. The complex chains of international production, highly integrated trade ties, and transnational social and business connections that characterize the California economy may also be responsible for this diminished importance of transportation costs. Above all else, we must also take into account the increasingly "weightless" nature of high-tech products, i.e. their high price-to-weight ratio.

The foreign-born immigrants settled in the country appear to constitute a formidable bank of competitive power, an observation shared by other social scientists. Further research is needed before any conclusive evidence is acquired regarding the role and impact of immigrants, and the role of these networks in the prospective "demise of distance". Disaggregation by traded commodity and by immigrant demographics, as well as further studies for other immigrant countries, such as Australia, might be able to converge on the underlying factors contributing to increased trade. Also, business networks set up by multinational firms, cross-border family enterprises, their subsidiaries, and affiliates, have a powerful and positive impact on overall exports from these sub-national regions, suggesting that there are spillover effects of foreign direct investment. The economic space created by the interaction of affiliates, subsidiaries and the parent firm tends to have an energizing effect on arm's length transactions as well.

Our attempt at the study of international trade of a sub-national region, together with the social and business networks that bind it to other national economies, is in effect a look at a regional economy as an integral part of the global. Literature on international trade has either ignored sub-national regions or treated them as fractal or self-similar to the national economy. At a time when national economies and global markets are becoming more closely integrated, a look inside the "black box" reveals the singular attributes of ties that bind different sub-national regions to the global economy.

Chapter 7

Global Linkages, the High-Tech Sector, and State Policy Choices

Our previous chapters have shown that the global forces affecting states and local areas are complex, with imported inputs and foreign direct investment abroad by local companies playing a parallel role to exports in economic growth and competitiveness. State policies toward global economic linkages almost always are directed to a much narrower set of global forces—export activity and support for businesses and workers at risk because of imports.

Individual US states often market their importance in terms of their economic size. The top two "Fast Facts" on California's Technology, Trade and Commerce Agency web page (2002b), for example, are "Fifth largest economy in the world," and "Gross State Product is $1.4 trillion." At least four other states—New York, Texas, Illinois and Florida—have GSP levels that would place them within the top twenty nations.[1] Yet each of the fifty United States faces a much more restricted set of policy options than a country of similar size. Export and import policies and trade agreements with different regions of the world are set at the national level. Immigration policies toward foreign nationals are also established at the national level.

As noted in Chapter 3, many business proprietors look towards the federal government to manage trade policy and to state and local governments for assistance in more general competitiveness factors. Nevertheless, recognizing the growing trends towards globalization of the US economy, many states are examining how policies at the state and local level may affect the competitiveness of their industries and their labor force in the global market

---

[1] *The World Factbook 2001* and US Bureau of Economic Analysis web page.

place. We look at these policies in this chapter and relate state trade programs and more general development programs to the characteristics of the global high-tech industry as described in this book.

7.1   FEDERAL TRADE POLICY FROM A STATE PERSPECTIVE

Both the structure of economic programs in the US and some broader policy debates influence the types of programs that have evolved and can evolve at the state level. The overall climate for trade has been influenced by changes at the national level. The 1990s brought legislation and trade agreements that led to increasingly free trade worldwide, and more specifically between the US and other countries. NAFTA (the North American Free Trade Agreement) established North America as a trading block, reducing barriers among the three partners—the United States, Canada and Mexico. The GATT "Uruguay Round" (General Agreement on Tariffs and Trade 1994) built on earlier global agreements to further reduce trade barriers at a global level and established the World Trade Organization (WTO). The same round of negotiations and subsequent agreements addressed new issues related to trade in services (GATS—General Agreement on Trade in Services) and to the information technology revolution (TRIPS—Trade Related Aspects of Intellectual Property).[2] Under GATT/WTO, "WTO rules now apply not only to the one-fifth of world production that is traded but also to goods and services that may never enter into trade,"[3] affecting sub-regions of nations whether or not they are heavily engaged in trade.

Immigration policy in the US also became less restrictive during the 1980s and 1990s, allowing more immigration and greater use of non-immigrant foreign labor than in many earlier periods. The number of immigrants admitted to the US rose from about 4.5 million (in total) during the 1970s to over 9 million during the 1991-2000 period, as shown in Figure 7-1. In 2000, 25.6% of immigrants admitted to the US were in or came to California. Under the non-immigrant foreign guest worker program (H-1B in its latest form), the number of visas granted rose from about 60,000 in 1990 to over 200,000 in 2001.[4] (See Figure 7-2). The future path of US immigration policy, however, is unclear due to the events of September 11, 2001 and the new security risks associated with immigration.

---

[2] See discussions of trade agreements, the WTO and challenges it faces in Deutsch and Speyer (2001), Rugman and Boyd (2001), Sampson (2001), and Zedillo (2003).

[3] Sampson (2001), page 3.

[4] California accounted for 16.2% of nonresident workers in 2000, according to US Immigration and Naturalization Service (2003), Table 41.

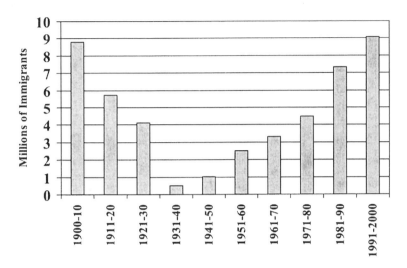

Figure 7-1. Immigration to the United States by Decade, 1900-2000. Source: Statistical Abstract of the United States 2002, Table 5. Note: Immigrants are "aliens admitted for legal permanent residency to the United States" and may include not only new arrivals who have been issued immigrant visas but also existing aliens residing in the United States who are changing their status from temporary to permanent residency.

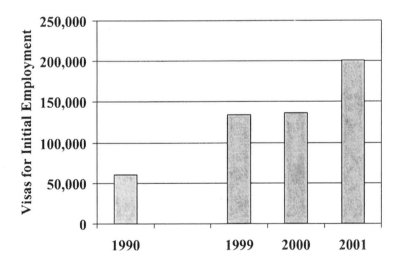

Figure 7-2. H1-B Visas Issued for Initial Employment, 1990 and 1999-2001. Source: US General Accounting Office (1992), US Immigration and Naturalization Service (2000) and (2002a) and (2002b).

A state influences national trade policy largely through its Congressional delegation. The California congressional representatives have, on balance, been supportive of the trend toward trade liberalization, as shown by voting records tracked by the California Council for International Trade (CCIT). A majority of California representatives voted in favor of ten out of fourteen of the trade liberalization bills tracked by CCIT between 1993 and 2000. However, support for trade liberalization is far from unanimous among California legislators. A significant minority—46%—supported less than half of the measures tracked by CCIT in the past decade.[5] California has shown even greater ambivalence toward expanding immigration, with high-tech corporations strongly supporting the H-1B program, but voters passing an initiative to deny services to undocumented immigrants.[6] Many of the H-1B visas are in professional and technical occupations which historically are not unionized. (See Figure 7-3.) Some professional organizations have questioned the necessity for the program, arguing that unemployed older engineers are a possible labor force to fill the gap in workers.[7]

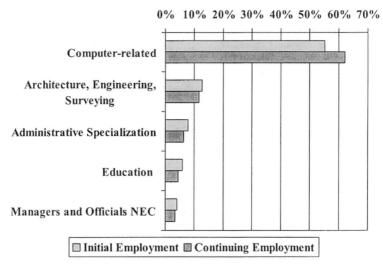

Figure 7-3. Major Occupations of H-1B Visa Holders, 2001. Source: US Immigration and Naturalization Service (2002a).

---

[5] The Policy section of www.ccit.net has a table reporting votes of House representatives.

[6] California voters passed an Initiative (put on the ballot by voter signature, not the state legislature), Proposition 187, in 1994, which sought to withhold education, medical and social services from undocumented immigrants. Court rulings later overturned much of the measure. See Mailman (1995) for a discussion of the measure and its early legal history.

[7] See Garry (1999), Braun (1998), Parks (1998), and American Association of Engineering Societies (2000).

7.2   IS THERE A PLACE FOR STATE PROGRAMS?

Overall support for trade comes from the perception that trade generates new business and job opportunities. Much of the ambivalence shown by the California delegation (and California voters) toward national trade policy is the recognition of the unequal distribution of benefits and costs of globalization. Key concerns have been the potential loss of jobs in industries going overseas, the loss of jobs in industries where H-1B immigrant workers are used, jobs displaced by low-cost imports, rising inequality of incomes, world-wide environmental effects of trade with countries with lower environmental standards, and public sector costs of services to undocumented immigrants. Some of these concerns are entirely out of the hands of state and local agencies (environmental impacts in foreign lands, for example), but other concerns—from changing industrial structure to state services—can be addressed at the state and local level.[8] A great deal of discussion in the academic literature as well as on the policy front has centered around how to provide effective responses to these concerns.

One ongoing dilemma has been whether to address the local problems that ensue from globalization at the national level, or at the state or local level. The academic literature is divided on the effectiveness of these approaches. Some of the neoclassical economic and public finance literature suggests that direct subsidies to individuals or firms who have suffered direct negative impacts from economic changes such as global restructuring are the most effective way of counteracting these impacts.[9]

A more recent critique of the effectiveness of economic development programs argues that local programs may help individuals but leave the region worse-off. For example, worker-training programs may help specific individuals, but have little impact on the region's overall unemployment rate. Alternatively, these programs may improve the region, but only by displacing the problem population or weaker industries. For example, the introduction of office structures in redevelopment areas have displaced lower value-added industrial activities such as printing and metal work, without necessarily providing new employment opportunities for those who were once employed in the displaced businesses. This has been a critique of US and UK redevelopment programs at various points in time.[10]

In contrast, the discipline of regional economics provides a rich body of literature that argues that place-based programs and industrial policy can be

---

[8] Concerns regarding the effects of trade agreements on state environmental policy are a separate matter, as discussed in Orbuch and Singer (1995).

[9] The general concept of improving the mobility of people is summarized in Hoover (1971), pages 258-259. Broader discussions are found in Winnick (1966) and Bolton (1992).

[10] Bovaird (1994).

very effective in counterbalancing the negative effects of the free market. Markusen (1994) refers to examples of successful regional policies in both Korea and the United States, but also describes the potential (and actual) conflicts between industrial and regional policies. Several authors describe the recent successes of local economic development programs that have moved away from smokestack chasing (which is often a zero sum game, moving jobs from one locale to another) towards capacity building "positive sum" efforts targeted to the characteristics of the local setting and industries.[11]

Most states make some effort to deal with the unequal impacts of trade. On the employment side, federal dollars available through the Trade Adjustment Act and other programs to help displaced workers allow states to set up employment training and other adjustment programs appropriate to the needs of displaced workers in the state's economy. Other training facilities, such as community colleges, may also be drawn on as resources for displaced workers, and in many states pubic/private partnerships are an integral element to training programs. On the industry side, small business development centers offer an array of programs available to help with planning, business management and finance. These programs are not necessarily designed for firms losing business to other venues, but are more generally aimed at assisting small businesses to grow or to recover from economic downturns.

Dealing with the negative aspects of trade is only one concern of state and local governments. The potential benefits from expanding the global role of local firms are often more prominent in the minds of state and local officials. States have a long history of programs specifically designed to take advantage of the perceived opportunities from the increasing levels of international trade and global production. In a survey conducted twenty years ago, Posner (1984) found that forty-seven of the fifty US states had some type of export marketing program, and the remaining three states provided other support such as education and counseling for firms interested in exporting their products.

The empirical academic literature has brought evidence in support of these programs. Several researchers have found that firms engaged in exports have higher productivity, higher growth rates, and lower failure rates than similar firms that are entirely dependent on domestic markets.[12] Researchers looking specifically at export programs have found correlations

---

[11] See, for example, Blair (1999) and Bradshaw and Blakely (1999). "Capacity" building implies increasing a region's productive capacity through strategies such as education or access to financing.

[12] A sample of these studies is summarized in Richardson, Feketekuty, Zhang and Rodriguez (1998).

between spending on these programs and levels of exports.[13] Later research showed a relationship between the level of state appropriations for exports and direct employment in export industries.[14] Causality remains a question in these studies, which do not determine whether existing levels of exports generate a need for the government services or instead, government programs help to stimulate an increase in exports and export-related employment.

This evidence suggests that despite the limitations and pitfalls of place-based policy approaches, there is a place for state and local policy in responding to the changes brought by globalization. However, state programs to deal with the opportunities and problems of global linkages do not exist in a vacuum. Many are put in place as part of a larger fabric of human resource and economic development programs designed to meet the needs of distressed communities and displaced workers throughout the state or to protect and promote competitiveness and augment opportunities for the state's business base.

7.3   THE ISSUES FOR STATE POLICY

Table 7-1 summarizes some of the advantages and disadvantages a state can experience from an increasingly global economy, as described in earlier chapters. A number of studies have shown that exports increase output and employment. A company that discovers strong demand in foreign markets, however, may soon decide it is more efficient to move production (and jobs) to a foreign location close to those markets. This investment abroad, while taking away jobs (or the potential for job growth) locally, may in fact lead to preservation of the remaining local jobs in an industry that otherwise would decline due to foreign competition. Competing imports that may lead to job cuts also provide cheaper goods for consumption and possibly less expensive inputs, improving the competitive position of companies using them. Foreign direct investment is often seen as another opportunity to increase a state's employment base, but the new firm may also be a competitor to the domestic firms already located in the state.

State and/or local programs generally will not address all of these aspects of expanding global linkages. It is thus important to evaluate the total impact of all policies, as well as of the individual policies one by one.

---

[13] Coughlin and Cartwright (1987).
[14] Wilkinson (1999).

Table 7-1. Potential Positive and Negative Effects of Expanding Global Interactions

Type of Global Interaction	Possible Positive Effects	Possible Negative Effects
Expanding State Exports	Add jobs, revenues to state businesses	Production may move abroad for successful foreign markets.
Expanding Direct Investment Abroad by US Firms	Repatriated profits for US multinational firms	Blue-collar and even technical jobs may move abroad
Import Competition	Lower cost goods	May reduce revenue and employment at local firms
Imported Inputs	Lower costs for US firms	Competition for domestic suppliers
Foreign Direct Investment in US	May employ US workers and buy from US suppliers	May compete with domestic firms in US markets.

Source: Authors based on research discussed in Chapters 2 through 6.

7.4 A CLOSER LOOK AT STATE "FOREIGN TRADE" PROGRAMS AND POLICIES

In most states, economic development policies are separated from labor force policies (unemployment and employment and training programs). Most programs specifically designed to deal with the opportunities of "foreign trade" are part of the state's economic development function, as are many programs for businesses in distress because of import competition or other causes of economic dislocation. Programs dealing with displaced workers are part of the state's broader function for serving unemployed workers.

Resources for foreign trade programs grew very rapidly in the 1980s and early 1990s. Between 1984 and 1992, "international appropriations" for all US states increased over four-fold, according to data from NASDA, the National Association of State Development Agencies.[15] Since that time, these appropriations have grown more slowly and even decreased. States budgeted a total of over $94 million for these programs in 1992 but only $76 million in 1998.[16] Despite recently stagnating or declining budgets, state programs for trade remain much more diverse than they were in the 1980s.

---

[15] Conway and Nothdurft (1996). "International appropriations" include export promotion, investment recruitment, and in some cases tourism, port and agriculture appropriations, according to the authors.

[16] These data are also from NASDA. The 1992 data are reported in Conway and Nothdurft (1996); the 1998 data are shown in National Governors Association (2002).

Figure 7-4 shows the types of trade-related programs found at the state level, as identified from each state's web site. The shift in emphasis has changed somewhat from the Posner survey of twenty years earlier, reflecting changing motivations, changing resources, and changing technology.

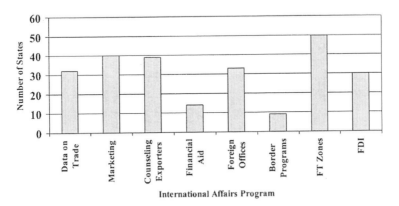

Figure 7-4. International Affairs Programs in 50 United States, 2002. Source: Authors from research on the World Wide Web, 2002. FT: Foreign Trade. FDI: Foreign Direct Investment.

7.4.1 Data on Levels and Trends in Trade

While most states in the 1980s had some type of printed material describing their trade programs and the state's economy, only a handful could provide detailed data on trade activity. Now two-thirds of states provide detailed statistical information on trade on their web sites, generally drawing from data provided by the US Department of Commerce Exporter Locator series or from revised MISER data.[17] Many states also do some types of market research, providing material that can be used to give advice to businesses seeking export assistance.

---

[17] These data series are discussed in more detail in Chapter 2. MISER data on California referred to in this chapter come from the State of California Technology, Trade and Commerce web site.

7.4.2 Export Development—Marketing and Other Promotional Activity

Export development is a frequently used state foreign trade policy. Development may include marketing products for export, providing information on potential trade partners to domestic firms, and providing services to facilitate exporting by domestic firms. Many states have export development programs, "most of which focus on helping smaller companies with limited resources."[18] The most common activities of these offices are promotion related, through trade shows, trade missions, and overseas trade offices. These programs are often criticized as ineffective, serving a small number of firms, and expensive if they involve overseas trade offices or missions.[19] A few programs are more strategically based; rather than trying to reach all firms, they focus on firm categories most likely to benefit from increased exports. Conway and Nothdurft (1996) cite Oregon's strategy of "focusing narrowly on finding overseas representation for the state's mid-market manufacturers and trade-service companies—those with $2-20 million in sales per year."[20] This is a strategy to target firms that are already well established but need new markets for expansion.

Despite their critics, trade shows, trade missions, and foreign trade offices continue to be a major use of resources. Web site information indicates that four-fifths of states are actively engaged in trade shows, missions, and other marketing activities, and two-thirds of states have at least one foreign trade office or contracted foreign representative. The presence of state offices has grown sharply over the past two decades, as shown in Figures 7-5 and 7-6, and the location of offices has shifted.

In the early1980s, 26 states had at least one foreign trade office. The total number of foreign trade offices grew significantly by 2002 (from 49 to 191), with a wide geographic distribution covering Asia, Europe and the Americas. In the realm of promotion, foreign trade offices may have several functions—helping firms to market their products abroad, identifying new market opportunities, and acting as the first point of contact for companies that may want to locate production in their state. In addition, foreign trade offices may have a "capacity building" role, to which we now turn.

---

[18] Scouton (1989).
[19] See, for example, Nothdurft (1992), Chapters 1 and 2.
[20] Conway and Nothdurft (1996), page 41.

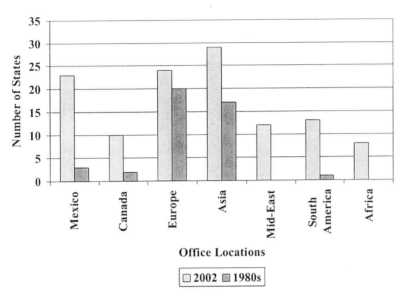

Figure 7-5. States with Overseas Office Locations, Early 1980s and 2002. Source: Posner (1984) and authors from research on the World Wide Web, 2002.

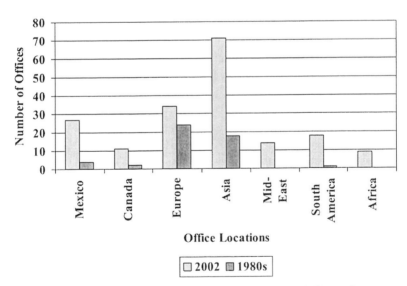

Figure 7-6. Number of State International Affairs Offices Overseas, 2002. Source: Posner (1984) and authors from research on the World Wide Web, 2002.

7.4.3   Export Development—Capacity Building and Finance

Some export development programs have gone beyond promotional activity. A number of states provide counseling to firms new to export activities and provide financial assistance for trade show attendance, promotional materials, and market research. Many states also recognize that other organizations—federal agencies (primarily the Department of Commerce and the Department of Agriculture), local government, and private trade organizations—are involved in similar export development, and design programs that involve cooperation with or coordination of overlapping activities. Contributing to these coordinated approaches are the network of Export Assistance Centers and District Export Centers established by the US Department of Commerce and involving the cooperation of state agencies, local agencies, ports, and some private organizations. The goal of these centers is to integrate resources for export promotion and finance from a variety of federal agencies as well as from state and local agencies and organizations, while tailoring the types of services provided to the economy of the region served.[21]

Similar integration may take place with respect to foreign trade offices, where a state shares representatives or resources with other states or with federal agencies. The foreign trade offices may help firms with marketing activity as well as with licensing requirements, local contacts, and networking new exporters with experienced firms.

7.4.4   Foreign Direct Investment at Home and Abroad

State efforts to encourage investment by foreign firms in their area (Foreign Direct Investment, or FDI) have often preceded efforts to encourage exporting by domestic firms.[22] Major activities include promotional programs, site selection and regulatory assistance, and foreign trade zones. In addition to the roles discussed above, trade offices, trade shows and missions may serve as a means of informing foreign firms of the opportunities for production activity within the US. For decades, most states also have acted as hosts to visiting trade missions from other countries. A smaller number of states have more extensive programs to attract foreign firms. These include marketing campaigns, site selection assistance, assistance with regulatory requirements, and tax and other types of incentives. South Carolina, Tennes-

---

[21] See discussion in "Small Business," *Business America* (1996), and in "US Export Assistance Centers: One Stop Shop for Exporters," *Business America* (1994). Updated information is available from the Small Business Administration web site.
[22] Nothdurft (1992), page 89.

see and Kansas all have prominent firm-recruitment web pages leading into their international affairs section (as one element of economic development). California, New York, Maryland and Oklahoma offer site location assistance, and some states include regulatory assistance, incentives, or "strategic alliances".[23]

State "international affairs" policies generally do not address the topic of investment by domestic firms abroad. However, states that assist firms in dealing with foreign markets, either through foreign trade offices or through designated geographic area experts in their within-state offices, may find that they are serving this need as well. The first step may come through assisting firms that are establishing foreign operations to promote sales and marketing efforts. Successful sales abroad may soon lead to the need for technical support establishments in the countries or international regions where the product is sold, again possibly with the assistance of state and federal offices. From this level of effort, the firm may expand to production abroad, suddenly raising the question of how far the state should go in contributing resources to assist in this effort.

7.4.5   Foreign Trade Zones—One Tool For FDI

The federal foreign trade zone (FTZ) program, which has been in place since 1934, is another tool that states may use as a location incentive for both domestic and foreign firms.[24] Among other advantages, the FTZ allows exporting firms to use imported inputs without paying tariffs, for the products that are then exported. Each of the fifty states has at least one foreign trade zone, and many have sub-zones (off-site facilities—often a single manufacturing facility—licensed to operate as a sub-zone of an existing FTZ).[25] A much smaller number of states explicitly market the zones to foreign companies, as one of the advantages of locating in the area.[26] For other economic development groups, the FTZ is used as a tool to help expand business of existing exporters or to retain firms that might otherwise locate part of their operations overseas.

The effectiveness of FTZs has been found to be mixed. On the one hand, individual firms can clearly benefit from the reduced-tariff advantages of an FTZ.[27] Some authors cite examples of foreign firms considering the avail-

---

[23] These programs are described on the individual state's web sites.
[24] Da Ponte Jr. (1997) provides an overview of the history of the program in the US.
[25] The Foreign-Trade Zone Resource Center web site lists all foreign trade zones in the US.
[26] Authors' review of web sites.
[27] A sample of articles outlining the advantages of locating in an FTZ include Anonymous (2002) and Krause (1993). Studies that have found benefits to these firms include Head, Ries and Swenson (1999) and Johansson (1997).

ability of an FTZ in making a US location decision.[28] From that point of view, a state with no FTZ would be at a disadvantage. On the other hand, a study of Japanese investments in the US between 1980 and 1992 found that while foreign trade zones influenced the location a firm chose within the US, these zones did not necessarily increase overall Japanese investment in the US.[29] A statistical analysis focusing specifically on the foreign-trade zone effect did not find an overall increase in exports as a result of the FTZ policy. The researchers concluded that the FTZ program would be effective only in the context of a much broader range of programs.[30] (The study did not address the question of whether FTZs helped US localities compete with sites worldwide that offer tariff and other incentives.)

7.4.6   Trade Displacement Assistance

The US Trade Adjustment Assistance program ensures that all states have some type of program for assisting workers displaced by imports. The programs are administered by state human resources agencies (as with the Employment Training Panel of the California Employment Development Department). The overall record of these programs is mixed. There is strong evidence that increased imports can lead to greater displacement of workers in an industry, when the imports replace domestic production (particularly of production workers, according to our own research reported in Chapter 4 above and in earlier studies).[31] There is little evidence that the trade displaced workers are out of work longer than workers who became unemployed for other structural reasons, or that the special training assistance provided (for trade displacement) helps shorten the duration of job loss or improves wages upon reemployment.[32]

More recent research on employment training and workforce development suggests the program features determine its degree of success in helping workers and communities. For example, research by Thomas and Ong (2002) found that technically qualified engineers who lost jobs during the contraction of the defense industry needed social training and education as well as related skills training to be able to compete for closely-related jobs in

---

[28] Rolfe, Ricks, Pointer and McCarthy (1993) found from a survey of firms that the tariff advantages of FTZs were a significant factor in attracting exporting firms to FTZ locations.

[29] Head, Ries and Swenson (1999).

[30] Knowles Mathur and Mathur (1997).

[31] See Bardhan and Howe (1998a), Bardhan and Howe (2001), and Kroll and Kirschenbaum (1998).

[32] Addison, Fox and Ruhm (1995), Decker and Corson (1995); it should be noted that these evaluations are not specific to the programs by the California Employment Training Panel.

electronics and computer industries. An article by Rosen (1998) also empha-
sizes the need for instilling flexibility as well as training in workers, advocating a "lifetime learning system to ensure that worker's skills are continu-
ally updated and enhanced."[33]

Less widely used are programs targeted to assisting businesses in industries at risk from import displacement. Without the ability at the state level to
offer protective tariffs, and without the resources or political support at the
state level to subsidize existing operations, state or local area programs to
assist import-threatened industries have sought other means of improving
local competitiveness. Strategies may include organizational assistance, new
product development, selective employment retraining, and production technology assistance. One example is the Wisconsin Regional Training Partnership, which addresses competitiveness in the Milwaukee manufacturing sector.[34] While originating through the actions of a governor-appointed panel,
the partnership's board includes business, labor and public sector leaders,
and programs draw on federal, state, local and private resources. Programs
are focused on modernization of local industry and improved competitiveness at the level of the production process and worker training. The "partnership" aspect has been key to the programs' success, with the participation of
managers and union leaders, as well as "neutral" third parties, such as local
technical colleges.

Manufacturing extension programs, which address the organizational as
well as training needs of firms, are another major element of the program.
Another example is Garment 2000, a business, labor and community college
consortium in San Francisco, which drew on a wide variety of public and
private, local and federal funding sources to improve conditions in the apparel industry while providing resources for modernization and workforce
training.[35] The program has become the main component of the San Francisco Center for Applied Technologies at City College of San Francisco.[36]

A third response to import displacement is the funding of appropriate
"transfer industries"--industries that have growth potential and could provide
comparable employment for workers displaced by foreign imports or by
other economic shifts that affect a whole industry. For example, California
has sponsored a number of advanced technology programs to provide long-
term opportunities for companies and workers to replace lost defense work.[37]

---

[33] Rosen (1998), page 84.
[34] Parker (1997).
[35] Chin (1995).
[36] See City College of San Francisco, Centers for Applied Competitive Technologies web site.
[37] See, for example, Koehler (1994).

7.4.7   Retention and Broader Strengthening of Competitiveness

Industries competing successfully in global markets, especially those with high trade flows as a share of output, are likely to be undergoing transformations in terms of employment structure and possibly firm structure and location. Whether firms move out of a given state as they undergo these transformations will depend partly on whether the state maintains or improves critical resources. The case study reported in Chapter 3 points to key competitiveness factors for the computer cluster. The great majority of interview respondents did not expect help from the state directly in terms of foreign trade. They showed much greater concerns with general resources (e.g. education) and operating conditions (e.g. environmental regulations and taxation) in the state. For the computer cluster, the state's higher education system provides resources for skilled labor as well as a framework of support for professional networks. A strong primary and secondary education system is important for the remaining production labor force and support staff, and also as a recruitment factor for attracting and retaining skilled labor. A parallel study of the food processing industry found that those firms required maintenance of quality agricultural land, university based product research, and a strong transportation system.[38]

A state such as California is particularly vulnerable with regard to retaining its existing large export industries. For many other localities within and outside the US, one strategy for expanding exports is the recruitment of strong exporting industries from California. Programs to identify and address the expansion needs of these industries, often on an individual firm basis, can be influential in determining whether expansion occurs within California or out of state. Other states use both retention and recruitment efforts. New York State, for example, has regional offices throughout the state that offer "one stop shopping" (site selection and technical assistance) for New York companies seeking new locations, as well as for firms from elsewhere in the US or abroad considering New York as a location.[39] Within California, Joint Venture Silicon Valley, a public-private partnership located in Santa Clara County, focuses heavily on competitiveness and retention of Silicon Valley Businesses. Task forces established by the organization address issues as diverse as improving education and skills of the local workforce, tax policy, affordable housing and transportation.[40]

---

[38] Kroll and Kirschenbaum (1998).

[39] *Business Assistance in the Empire State* (promotional material published by Empire State Development, New York State's economic development organization).

[40] The Joint Venture Silicon Valley web site provides detailed information on the groups programs and strategies.

7.4.8   A Multi-Jurisdictional Approach

The issues raised by international trade at the state level generally cannot be effectively addressed by programs that operate only at the state level. Much of the jurisdictional power influencing industry-wide trade is, of course, at the federal level. State policy-makers will wish to understand the implications of proposed or newly enacted trade agreements for key sectors of the state's economy, recognizing differential effects either among industries within the state or among firms or wage earners within an industry. With this understanding, federal agencies can become an important resource for trade related programs or for programs that would mitigate negative impacts. Particularly in the areas of export promotion, export finance and trade adjustment assistance, state programs can build on and help direct firms toward the use of federal resources.

At the other extreme are regional and local organizations that provide trade-related assistance or broader economic development strategies. An important role of state international trade organizations may be to encourage the collaboration among jurisdictions and between private industry, educational institutions, and public and private development organizations in providing resources to firms affected (positively or negatively) by foreign trade.

While working cooperatively with both federal and local level resources, there are also distinct activities that can be best directed from the state level. In the area of export promotion, the state can play a role as a focal point for exporters, especially in industries that are not already served by federal programs or by strong trade organizations. In the area of labor force preparedness, the state has a critical role in providing resources for training, in supporting the public school system, and in supporting higher education. In the area of firm expansion and location decisions, the state can help identify the critical location factors for key industries and provide support and mediation for firms ready to expand or relocate.

Many of the state-level activities mentioned here in a foreign-trade context are of equal importance to a state's overall economic development strategy. Because industries with strong foreign trade components are also likely to be undergoing change in the location of production and occupational composition, understanding the ramifications of an industry's trade component may help state agencies to use existing economic development resources more effectively.

7.5　　RESOURCES IN PLACE: A CALIFORNIA CASE STUDY

California has a number of programs in place to address some of the opportunities and impacts of foreign trade discussed here. The state's trade-related programs can be roughly divided into three types: those related directly to foreign trade, those that focus more broadly on competitiveness, and programs that are concerned primarily with firm retention.[41] These are summarized in Tables 7-2 and 7-3.

Table 7-2. California Programs Directly Related to Foreign Trade

Program Name	Agency	Purpose/Activities
International Trade and Investment Division	Technology, Trade and Commerce	Trade delegations, missions, seminars, promotional tools to assist California companies in foreign market transactions
California Export Finance Office	Technology, Trade and Commerce	Loan guarantees for export business, small and medium-sized companies
Office of Export Development	Technology, Trade and Commerce	Marketing services for small and medium-sized California companies involved in exporting
Division of Tourism	Technology, Trade and Commerce	Tourism information for both domestic and foreign tourists
Office of Foreign Investment	Technology, Trade and Commerce	Assists foreign firms seeking to locate in California with site location, regulations, incentives, and information
Agricultural Export Program	Food and Agriculture	Assists small and medium-sized companies to expand their export activities.
Energy Technology Export Program	Energy Commission	Seed money, market and trade analyses, trade missions; facilitates California firm interactions with foreign energy officials
Foreign Trade Zones	Regional operators/grantees	16 zones provide tariff reductions and postponements for manufacturing and distribution firms
Source: Koehler (1994) and (1999), Koehler and Hogan (1996)		

---

[41] A much more detailed discussion of California trade programs is provided by Koehler (1999).

7.5.1 Programs Directly Related to Foreign Trade

As of 1998, the most recent year for which state-by-state comparative data are available, California had the largest state budget for international affairs programs. The $8.98 million budget was over 70% above the next highest (Pennsylvania). These impressive figures, however, are quite misleading. The economic development function is, on average, less than 0.25% of a state's total expenditure.[42] In 1998, California spent only 0.04% of its budget on the agency in charge of economic development programs.[43] International affairs programs are usually a component of the economic development program, and on average account for only 0.006% of state expenditures—California's share was only 0.0044% in 1998.[44]

During the healthy economic years of the late 1990s, economic development and international affairs budgets rose in California. The Technology, Trade and Commerce Agency budget rose to over $200 million by fiscal year 2001-2002, of which about $12.5 million was allocated to International Trade and Investment. Within International Trade and Investment, the budget is divided among several operations, with about half of the budget going to the state's foreign trade offices.[45]

The International Trade and Investment Division of the Trade and Commerce Agency is the state's primary vehicle for export development and promotion. It sends trade delegations overseas; through its foreign offices, it helps firms identify markets or distributors for their products; and its Office of Export Finance provides loan guarantees for exporters moving into overseas markets. As is typical of state-agency programs engaged in such efforts, they are used by a relatively small number of firms but can cite significant growth effects for those firms.[46] (This is consistent with reports from our interviews that indicated that few firms used these programs.)

Although not necessarily well known to their potential constituents, California's foreign offices tend to be well aware of firms with international interests in their areas. A 1999 listing of California firms operating in foreign

---

[42] The NASDA surveys from which these data are drawn use a narrow definition of economic development, focusing on the agency exclusively devoted to economic promotion, strategic planning, and project development. As pointed out by Riches McCullough and Ross (2002), Economic development more broadly defined includes employment training, education, and infrastructure and may involve billions of dollars in multiple agencies, and a much larger share of the total budget.

[43] The definition of economic development is quite fluid. When transportation infrastructure improvements, training at community colleges, or unemployment insurance programs are included in the estimate of costs, the budget share rises sharply.

[44] Authors' calculations from NASDA data, National Governors Association (2002).

[45] California Department of Finance (2002a) and (2002b).

[46] See Conway and Nothdurft (1996), pages 44-49.

locations overseen by the offices includes about 900 companies at almost 2,000 separate sites.[47] Consistent with the strong export (and global sales) orientation of high-tech sectors, high-tech manufacturing firms accounted for 37% of the sites and one-third of the companies. The state's large computer and semiconductor firms were prominent on the list (eg. Hewlett Packard, Cisco, 3Com, Apple, Intel). Information and software firms accounted for 13% of sites and 14% of companies listed, and included large software and networking firms such as Autodesk and Oracle.

The program's fiscal report for 1999/2000 divides the outcomes of their efforts by "Stateside" and "Overseas" offices. Stateside offices (California Export Finance Office, Office of Export Development, and Office of Foreign Investment) reportedly assisted in 70 deals, valued at over $3 billion, and creating over 6,300 jobs.[48] Over 6,000 of these were from seven deals assisted by the Office of Foreign Investment, bringing foreign firms into the state.[49] Overseas offices reported assisting 130 firms in deals valued at about $250 million and creating about 1,900 jobs. The total number of jobs created is tiny compared to the state's employment base of 15 million, but is in scale with the share of the state budget allocated to these activities. The numbers do not include business generated by more broad based promotional activities that do not lead directly to a deal but may stimulate interest in California products. The report also does not make it possible to determine the importance of the role played by these offices in the final deals.

California's Division of Tourism is also housed in the Technology Trade and Commerce Agency. The agency develops promotional material on California, markets tourism in the state, provides informational material, and conducts research. The division works cooperatively on promotional efforts with local government agencies and tourism-related organizations. The budget of the Division of Tourism was $8.5 million in FY 2001-2002. While the great majority of travelers in California are state residents (85%), foreign visitors account for almost one fourth of nonresident visitors.[50] The state Division of Tourism estimates that California received $11.3 billion in international expenditures from tourism in 2000.[51]

---

[47] Authors from data provided by the California Trade, Technology and Commerce Agency.

[48] A summary of the division's activities is provided in California Technology Trade and Commerce (2001). It is not possible to use these statistics to determine how many of the jobs would have been developed without state trade office assistance.

[49] As noted in earlier discussion and in Table 7-1, these moves are generated by a desire on the part of foreign producers to gain better access to California and US markets, and may to some degree bring in competitors to California producers.

[50] Estimates from statistics reported at the Tourism Division web site, statistics section of gocalif.ca.gov/state/tourism.

[51] See Tourism Division web site, Research and Statistics section. For more discussion on these types of estimates see Kroll and Bardhan (1996).

Table 7-3. California Competitiveness and Retention Programs

Program Name	Agency	Purpose/Activities
California Economic Strategy Panel	TTCA*, EDD*	Biennial strategic planning based on California economic regions, industry clusters, and cross-regional issues
Employment Training Panel	EDD	Funds training for new and existing employees, to assist businesses in obtaining skilled workers
Office of Major Corporate Projects	TTCA	Site location assistance and business advocacy for California corporations.
Team California	TTCA	Network of economic development professionals for business attraction, retention, expansion, job creation.
Red Team	TTCA in partnership with governor and other offices	Task force to address a particular issue or solve a dilemma to "keep California business competitive"
California Infrastructure and Economic Development Bank	TTCA	Low-cost financing to public agencies for infrastructure and public improvements (revolving fund); revenue bond program for private sector companies
Enterprise Zones	TTCA	Tax incentives to encourage business investment and job creation in economically distressed areas
Small Business Development Centers	State and US SBA* partnership	One-stop access for business counseling, planning, marketing and training programs
State Research and Technology Programs	CCST*, CAL-TRANS, Air Resources Board	Encourage the development of new technologies in which California could have a competitive edge.
California Higher Education Facilities	University of California; California State University	Higher education, basic and applied research
K-12 Initiatives	Governor's office, State Dept. of Education, public schools	Funding for school construction and modernization, class-size reduction, computer facilities
Centers for Applied Competitive Technology	Community Colleges	Manufacturing business improvement centers; advice and services, small and mid-sized firms
California Manufacturing Technology Center	Non-Profit outgrowth of joint state/federal effort	Improve methods of management and manufacturing in small and mid-sized businesses

*TTCA: Technology, Trade and Commerce Agency; EDD: Employment Development Department; CCST: California Council on Science and Technology; SBA: Small Business Administration. Source: Koehler (1994) and (1999), Koehler and Hogan (1996); State of California Web pages.

7.5.2   Competitiveness and Retention Programs

Table 7-3 lists a wide range of other California programs that affect the competitiveness of the state's computer cluster and other businesses. The list primarily includes resources in employment support, education, and economic development, but also touches on the roles of other state agencies such as transportation.

Over the past decade, California has taken an aggressive stance towards improving the competitiveness of the state for attracting new businesses and retaining existing businesses. Efforts have included strategic planning for economic development; workforce-related efforts focused on both short-term training and long term investments; programs directed at improving the operations and market penetration for individual firms; and infrastructure improvements that affect the population and businesses more broadly.

7.5.2.1   Strategic Planning

Since 1993, the state's Economic Strategy Panel (ESP) has overseen and funded analysis of the state's economy, identifying key growth sectors and critical issues facing these sectors. Beginning with a series of working papers and an interim report published in 1994, the ESP outlined an approach for the state that involved shaping programs to the needs of key industrial clusters and coordinating efforts among federal, state and local organizations. The ESP has also worked with existing regional organizations within the state and helped to establish new organizations, where necessary, to develop regional partnerships and programs to support the retention or expansion of key growth sectors.[52] The panel regularly updates their efforts approximately every two years.[53]

7.5.2.2   Education and Training

Education and training is among the areas recognized as critical by the ESP, and a number of programs exist to support this continuing need. Perhaps the most effective long-term investment in economic development ever made in the state was the higher education system. The need to maintain this system is one factor widely recognized by state, business and local strategists.[54] K-12 education is probably the area of most concern, in terms of the current level of programs and results. A number of programs have been established to address concerns with the K-12 system, ranging from state funding of class-size reductions; to supplying funds for facility construction and

---

[52] California Economic Strategy Panel (1996).
[53] See California Economic Strategy Panel (1998), (2000a) and (2000b).
[54] California Economic Strategy Panel (1996).

modernization and for computers; to establishing performance standards and flexible programs for meeting standards.[55] In the area of skills training, the state Employment Training Panel has resources to fund training for new and existing employees, with the goal of providing skilled workers for business.[56] Some of the funding is directly targeted to workers displaced by import competition, but much of the funding more generally applies to workers in transition because of any type of competition—domestic or foreign—or because of technological advancements in the workplace.[57]

7.5.2.3   Business Development

The state participates in a number of multi-jurisdictional programs to improve the operations of manufacturing firms and other businesses or to encourage the development of new industries in California. State level participation comes from the Trade and Commerce Agency, the university and community college systems, and the Employment Training Panel. Local development organizations and industry consortia also participate in some of these efforts. Programs include small business development centers (a cooperative effort with the US Small Business Administration),[58] manufacturing outreach centers (operating through the community college system, offering new technology to businesses and training for employees),[59] manufacturing technology centers (federal and state funded to improve technology and management),[60] and specialized industry development programs such as CALSTART.[61] Many of the programs are small business oriented.

7.5.2.4   Infrastructure Improvement

Heavy investments in infrastructure allowed California to develop the economy it has today, but as with its education system, maintaining physical infrastructure has become an issue. The California Commission on Building for the 21st Century has done an initial assessment of the broad infrastructure needs of the state, from housing to transportation. The California Infrastructure and Economic Development Bank provides a revolving loan fund for public agencies and revenue bonds for business projects.[62]

---

[55] California Dept. of Finance web page (August 1997).
[56] Koehler (1994, page 19).
[57] Employment Training Panel web site.
[58] Description and links provided at Technology, Trade and Commerce Agency web site.
[59] Described at http://www.cact.org/.
[60] Now evolved into a nonprofit consulting group, partially funded by state and federal programs; see http://www.cmtc.com/CorporateInformation/Background.php.
[61] A description of CALSTART is available at www.calstart.org.
[62] Technology, Trade and Commerce web pages.

7.5.2.5   Business Retention

Business retention became a focus of state policy with the 1991-1993 recession. Substantial changes have been made with regard to some concerns, such as Worker's Compensation legislation. Other concerns, such as tax reform, have generated greater controversy because of revenue shortfalls during the recession and budgetary demands in areas like education. Simplification of the state's environmental regulatory structure (complicated by local control of many development decisions) has remained a goal, but no policy changes have ensued. Any changes in this area are also controversial, because of strong support for environmental quality among Californians.

In the absence of broad changes to environmental regulation, the governor's office under Governor Wilson established "TeamCalifornia," which has now become a part of the Technology, Trade and Commerce Agency. TeamCalifornia acts as a network of economic development agencies and functions dealing with business attraction and retention. Several programs exist to help businesses work through the challenging regulatory maze involved in locating and operating in California. The Office of Major Corporate Projects offers site location and business advocacy. In critical cases, TTCA in partnership with the governor's office and other agencies sets up "Red Teams" to assist firms on an ad-hoc basis, as needed, when they are considering expansion or relocation from a California site or are facing other regulatory problems which might later lead to relocation. [63] This approach was cited by some of the respondents in the computer cluster as being instrumental in keeping their production facilities in the state.

7.6   ISSUES IN STATE TRADE POLICY

California is probably typical in having a small program specifically targeted to trade issues embedded in a much larger set of economic development and human resources programs that affect trade or its impacts in one way or another and are spread among many different agencies and organizations throughout the state. Two issues arise because of the size and complexity of state trade services. First, coordination and focus become critical to making state trade policy effective. Second, because many trade-related programs are scattered and serve small constituencies individually, these programs become vulnerable in times of economic weakness, when budgets are tight.

Several studies have raised concerns about the need for coordination and focused goals in California economic development services in general and

---

[63] See Technology, Trade and Commerce web site for further descriptions of these offices and programs.

foreign trade programs in particular. In *Maximizing Returns*, the California Budget Project describes the state's lack of a structural framework for economic development spending, which the report found was spread among more than two dozen organizations and six dozen tax expenditures.[64] A 2001 evaluation by the California State Auditor of the core economic development agency, the California Technology, Trade and Commerce Agency, saw it as an agency with satisfied customers but one that was weak in coordination of activities with other activities and in leveraging resources by this coordination.[65] The evaluation of foreign trade activity, in particular, was mixed, with effective coordination identified among the Office of Foreign Investment, foreign trade offices, and local economic development agencies, but much weaker use of other entities for Export Development and Export Finance offices. A 1999 report by Koehler takes a broader look at state trade policy and points to the potential for developing "a comprehensive economic trade development strategy" modeled after the Economic Strategy Panel approach, focusing on regions and industries and the resources available throughout the state.[66]

All of these studies emphasize the challenge of coordination and the potential rewards of building a strategic trade or economic development policy that more explicitly identifies goals and objectives, and identifies roles and expected outcomes for the myriad of state programs that could be involved. This type of approach would allow for much clearer evaluations of the effectiveness of different policies. In the absence of any measures of effectiveness, the programs become vulnerable in times of financial stress.

The 2002-2003 budget process shows how vulnerable economic development and foreign trade programs are in the face of tight budgets. The Technology, Trade and Commerce Agency, for example, received a 24% cut in its proposed funding from the state General Fund, leaving the general fund portion of the budget 21% below the level for the previous fiscal year.[67] Some foreign trade programs were affected even more strongly. Funding allocated to foreign trade offices was cut by one-third from both proposed and actual levels, and the Office of Foreign Investment lost a similar share of funding. The export development office, which was already budgeted for a 16% cut from the previous year, lost an additional 25% of its budget.[68] Programs built on federal funding may find themselves similarly vulnerable.

---

[64] Riches, McCullough and Ross (2002).

[65] California Bureau of State Audits (2001).

[66] Koehler (1999).

[67] The California Infrastructure and Development Bank accounts for two thirds of the Technology, Trade and Commerce Agency funding for 2002-2003, with funds already allocated from bond measures.

[68] California Department of Finance (2002a) and (2002b).

The California Manufacturing Technology Center, for example, is one program built on the US Manufacturing Extension Partnership program, which was facing a proposed 88% cut in funding in FY 2003.[69]

The effectiveness of these programs is often not at issue when the cuts occur. The state programs drawing foreign investment into California and the US Manufacturing Extension Partnership are two programs that are reputed to bring in public revenues through tax payments and multiplier effects on the general economy far exceeding the program expenditures. However, a strong record of effectiveness may not be enough to keep a program going if the returns do not revert directly back to the agency and if there is no overall strategic plan justifying the program.

7.7   A COMPREHENSIVE VIEW OF STATE TRADE POLICY

Overall, foreign trade and globalization have brought many benefits to California's economy and to the computer cluster, as discussed in earlier chapters of this book. The state has proved to be well positioned to take advantage of the increasing global opportunities, and many of the changes in the labor force and in the general population have been in concert with these trends. At the same time, the opportunities have brought with them new issues, including instability of employment, fluidity within jobs, requiring workers to be flexible and to regularly learn new skills, widening inequality of incomes, and the displacement of some workers.

Many of the issues that arise from foreign trade also relate to broader economic development issues for the state—how to retain a portion of production of existing firms as they expand, assist displaced workers as some work moves overseas, and train a labor force for the growing economic sectors. California, like most other states with a significant high-tech presence, has a base of existing programs from which these needs can be addressed. The many different implications of expanding global linkages should be considered in setting priorities for these programs or establishing new programs.

What remains is for state policy makers to develop a stronger vision of how to meet the changing needs generated by growing globalization in a framework that also addresses other challenges facing the state's economy. At this stage, some programs are clearly part of a coordinated strategy, while others appear to be individual initiatives, not linked to broader programs or

---

[69] Bailey (2002) and NIST web site. Note that NIST, the agency that houses the Manufacturing Extension Partnerships, sees this cut as part of the long term plan to make the partnership program self-supporting.

goals. The characteristics of the global California economy described in earlier chapters would be consistent with a broader multidimensional view of trade-related issues. Several factors are key to creating this vision:

- *Identifying Needs and Setting Priorities:* The needs created by globalization are part of a broader set of demands facing California as the economy develops. Programs to promote exports, increase the competitiveness of firms facing import competition, and retrain displaced workers should be developed in the context of the broader economic development needs of the state.
- *Recognizing the Complex Effects of Global Linkages:* Taking advantage of global linkages is not just a matter of helping California firms to expand exports and foreign firms to locate production in California. Exporters may soon become producers abroad, while foreign investors in California may soon become competitors in markets served by California firms. As the California economy transforms in response to global pressure, it should be a state goal to ensure that companies and workers have the resources and training that will enable them to prosper in the changing economy.
- *Anticipating the Effects of Change and the Needs of Industries:* A good understanding of how industries operate and what are their major concerns should enable state and local government to anticipate and address potential issues before they become "push" factors, leading firms to seek out of state sites. Overall, the state approach to expanding foreign trade, and to economic development more broadly, will be most effective if it is proactive. As one respondent in our interviews commented, "The State needs to identify problems before they happen. Instead, state programs only start working when problems arise." Careful monitoring and analysis of successful industries by state agencies or state-supported programs could identify long term concerns (such as capacity constraints in Silicon Valley and the search for new growing sites of expanding agricultural sectors such as wine) and allow the opportunity for strategic governmental responses in anticipation of these concerns.
- *Identifying and Nurturing New Locations for Expanding California Clusters:* In addition to the individual case approach of the Red Team program, helping industries deal with expansion and congestion issues within the regions where they already concentrate, the state could work with businesses and communities to identify alternative regions within California where expanding industry could locate. For example, the expansion of a new University of California campus in the Central Valley with an emphasis on information technology could be coordinated with the expansion needs of growing high-tech businesses. (Meeting the ex-

pansion needs of one industry, however, may at times come into conflict with the land use needs of a different industry, such as with agriculture in the Central Valley).

- *Including Adjustment Programs as a Key Element of Trade Policy:* Trade clearly causes adjustments for specific industries and for occupations within industries. State programs can take advantage of resources ranging from educational institutions to facilities available at military bases to assist firms, industry groups, and individual employees in adjusting to trade-induced changes.

- *Developing Programs in a Multi-Jurisdictional Context:* While there are many unmet needs, there are also programs that appear duplicative, especially across government levels. The state could potentially play a strong role in coordinating resources at different government levels and in establishing networks of agencies that can meet different trade-related needs.

- *Including Monitoring and Evaluation as Part of Each Program Design:* It is important to track the changes affecting the state as a whole (e.g. increases and decreases in export activities), the changes experienced by industries or occupational groups within the state, and the activities of different programs, and to evaluate regularly whether the programs are providing appropriate services where the need is greatest. This can be best accomplished if general trends are regularly reported and if programs are required to keep careful accounting of their activities and services.

- *Targeting Programs and Sharing the Costs of Trade Assistance:* Overseas foreign trade offices and missions are among the more costly economic development programs engaged in by the state. State resources will go further if the programs concentrate on businesses with a history of successful production for other markets and if at least a portion of the cost is recouped from successful clients. Furthermore, this is a program area where evaluation could be particularly helpful in determining which types of firms or sectors can benefit most from the assistance of foreign trade offices.

California is fortunate in having many economic development resources already in place and in having an economy that has shown resiliency in its ability to recover from severe structural changes. From the point of view of foreign trade, what remains is to balance the resources to meet the needs generated by increasing trade and opportunities that move in both directions, into and out of California, and have differential effects on industries and workers within the state. What is needed is a strategy that goes beyond narrow focus of exports, recognizing the importance of global links in both directions, and linking other crucial programs that cover a wide range of needs, from location, land use, and housing related concerns to education, training, and related retraining needs.

7.8 POLICY IMPLICATIONS FOR THE HIGH-TECH CLUSTER

The discussion in the preceding section relates to state policy with regard to trade in any globalizing industrial sector. The conclusions from earlier chapters in the book suggest some additional implications for policy directly relevant to the computer cluster and related high-tech sectors. Most of these implications are relevant to any US state with a significant high-tech base.

- *Global Linkages Beyond Exports:* For US high-tech firms, in addition to exports, global linkages may involve overseas production as well as the imports of inputs. Links established by foreign trade offices for export promotion may be helpful as well in these more controversial activities. The research in this book suggests that these types of overseas linkages are crucial to the health of firms and to their continued location in high cost areas such as California. Were foreign trade offices alerted to this, they could provide assistance to California firms seeking to improve overseas linkages and at the same time would be able to identify industrial sectors that are at risk of losing business to overseas competitors and could identify how to respond to these changes in production patterns.
- *Role of Immigrants:* The immigrant network has contributed to the high-tech cluster by contributing to the knowledge base of the labor force and by fostering business links overseas. Policies with relation to immigration need to recognize the positive benefits of these networks.
- *Location and Congestion:* Interviews with high-tech business executives suggest that the pull of high-tech centers such as Silicon Valley is strong, despite the cost and congestion disadvantages of these locations. It may be possible to counter some of the pull from overseas by encouraging growth centers in close proximity to existing high-tech centers, but where costs and congestion are lower.
- *Systems to Respond to Industries Under Threat:* Our research has demonstrated that while these industries are always innovating and generating new growth, they also go through regular cycles in which production processes—and the location of production—change. This is accompanied by business failures and job displacement. The disruptions of this process could be reduced by an early warning program that identifies industries or portions of industries at risk and applies business assistance

and employee retraining programs proactively, before displacement occurs.

- *Shift to Services:* The shift away from manufacturing to services has been very strong in the high-tech sector in California. In the 1990s, this was accompanied by a rise in services, professional and administrative employment that more than compensated for the loss in manufacturing jobs. However, the long-term implications of this trend are not well understood. There may well be an advantage to maintaining a base level of manufacturing activity within the state, both fostered by and in order to contribute to, the process of innovation. Many industrial sub-sectors are under threat and will not retain a significant presence in the state in the not too distant future. These sub-sectors can be identified from the similarity of the occupational and skill mix of the people employed in them, as well as by the similarity of their input structure to that of those sectors that have already been adversely affected. It has to be recognized that constant innovation leading to creation of higher value added jobs is the only way for the state to retain its competitive edge.

- *Services Outsourcing:* Outsourcing of services has strengthened in the past five years, and this process will continue, threatening the jobs that have been growing even as manufacturing jobs decline. Outsourcing of computer software and services is only one piece of a larger trend towards foreign outsourcing of office occupations ranging from payroll processing to legal research. As higher level business services, professional and administrative jobs move overseas, the responses outlined for manufacturing, such as strengthening education and training and supporting innovation, will become equally applicable to nonmanufacturing activities.

Chapter 8

Conclusions

This book has studied the changing global linkages and patterns of production of a high-tech economy and the implications of these forces for the region where the industry is concentrated. An understanding of these transformations requires a redefining and broadening of the concept of globalization beyond just international trade. The past decade has seen dramatic changes in the international division of labor, particularly in the global distribution of high-tech manufacturing production. A similar process is now underway in the high-tech services industry, including computer software and business process outsourcing services. This chapter summarizes the key findings of the book relating to the globalization of a high-tech regional economy, including structural changes in high-tech manufacturing and non-manufacturing activity, regional effects, and policy concerns.

8.1   SHIFTING PATTERNS OF HIGH-TECH TRADE

Several key factors characterize the international activity of the US high-tech sector, as it has evolved over the past decade:

- The US trade deficit in high-tech hardware has continued to expand, while the US has maintained a large trade surplus in high-tech services.
- Imported goods represent about 50% of the US high-tech hardware market. A large part of these imports are imported inputs used in the production and assembly of high-tech products within the US.

- The import share of total inputs used in overall US manufacturing has shown a steady increase throughout the period from 1987 to 2000.
- A significant proportion of US trade is in the form of intra-firm trade. About 52% of US imports in 1997, for example, arrived through intra-firm trade (the remainder represented arm's length transactions).
- Imported inputs represent an increasing share of total US intra-firm trade. In 1992, most imports of US intermediate goods were the result of arms-length trade. By 1997, intra-firm trade had become a key source of imported intermediate inputs for US industries.
- US multinationals were responsible for more than two-thirds of all imports of high-tech intermediate inputs into the US.
- Transportation costs represent little hindrance to high-tech intermediate imports, consistent with their high-value, low-weight, attributes.

8.2   LINKAGES EXPLAINING REGIONAL EXPORTS

Our analysis of the high-tech exports of a regional economy, specifically California, leads us to the following conclusions:

- California high-tech exports represent over 50% of the state's total exports and over 38% of total US high-tech exports. Trade flows play a particularly strong role in California high-tech manufacturing sectors, relative to comparable US sectors and also relative to other California manufacturing sectors.
- A gravity model provides a useful first step for describing the export patterns of both the US and California's regional economy.
- California and the rest of the US share similar overall patterns for export destinations, although California exhibits a greater propensity for exports to the Asia-Pacific region.
- A large population of foreign-born immigrants tends to raise overall exports, and to direct these exports to their home countries.
- Transportation costs, proxied by distance, matter less for California's exports when the transnational networks of immigrants are taken into account. This is also related to the 'weightless" nature of high-tech exports.
- Business networks operating through multinational firms have a powerful and positive impact on overall exports from California and the rest of US, suggesting that there are spillover effects of foreign direct investment. The economic space created by the interaction of affiliates, subsidiaries, and the parent firm tends to expand arm's length transactions as well.

8.3 GLOBALIZATION AND STRUCTURAL CHANGES

The effects of globalization on a high-tech, regional, economy include changes in the location of production, structural impacts on the region's labor force, and changes in the mix of employment within manufacturing and between manufacturing and services:

- Despite strong growth in high-tech sales, the US lost more than 150,000 high-tech manufacturing jobs between 1990 and 2002, while adding almost 1.5 million high-tech service jobs. By 2002, the ratio of high-tech service jobs to high-tech manufacturing jobs exceeded three to one.
- Between 1987 and 1992, globalization in the form of foreign outsourcing accounted for one-third to one-half of the increase in relative inequality between blue- and white-collar workers in California.
- Industries experiencing sharper sales declines were more likely to restructure their production processes by substituting manufactured imported intermediate inputs for domestic blue-collar labor. Foreign outsourcing was thus a form of recessionary restructuring.
- Value added and value added per employee in high-tech industries have increased significantly, both in the US and in California. Foreign outsourcing positively impacts value addition, and may have thus played a key role in the increased profitability of US high-tech firms and in the run-up in their stock prices in the 1990s.
- Although the share of blue-collar wages in total payroll increased during the boom period 1992-1997, the source was new production jobs created by innovation in Silicon Valley. This trend has likely reversed with the post-2000 recession.
- Outsourcing became increasingly important in software, computer services, and other high-tech related services sectors in the second half of the 1990s. Some forms of outsourcing appear to have further accelerated during the post-2000 downturn in the technology industry.

8.4 REGIONAL CONSEQUENCES OF GLOBAL HIGH-TECH: THE CALIFORNIA CASE STUDY

High-tech globalization and structural change have affected the composition of employment, trade flows, and competitive pressures in high-cost markets like California:

- California location quotients confirm the importance of both high-tech manufacturing and services for California's economy. The coefficients for material inputs are particularly high, showing the importance of component outsourcing for the production process.

- A significant part of the growing global demand for products from California companies will likely be served from the foreign subsidiaries or affiliates of California multinational enterprises, rather than by direct California exports. Only a small proportion of the direct job growth (but perhaps a larger share of income growth) will occur in California.

- Globalization is a factor leading to the restructuring of production and jobs in the California computer cluster, in particular to the shift from manufacturing and production jobs to services jobs and non-production jobs within and beyond manufacturing.

- Increasing globalization affects the employment mix in California's computer cluster, keeping high-paying and technically demanding jobs in California, while moving more routine jobs overseas.

- Recent global expansion of the outsourcing of service activities, such as software production to India, Russia, and China, raises the prospect that a significant share of California's high-tech service sector employment may be lost to foreign outsourcing, following the same pattern as the earlier loss of high-tech manufacturing jobs.

- Imported intermediate inputs and overseas production have complex effects on California's economy. While California loses manufacturing jobs as a result of these forces, the reduced costs improve California firms' competitiveness and allow them to expand world-wide, thus increasing high value-added activities for the state.

- California's continued competitiveness relies on the maintenance of its underlying advantages. For the computer cluster, this includes the existence of centers of high-tech activity, skilled labor, institutes of higher education, social and cultural dynamism, openness, environmental quality, and other "quality of life" variables.

8.5   STATE POLICY ROLE

A state such as California, with an economy larger than that of most nations, seeks to influence the ways in which the international economy and global linkages have impacts on the state's businesses and labor force:

- Companies look primarily to the federal government for assistance in trade matters, while they look primarily to state and local governments to improve the region's competitiveness.

- A state's global strategy should look beyond simple export promotion, to consider the global needs of key industries, which may include critical imports or reliable production networks abroad.

- A state strategy to respond to the changing structure of global production must proactively address elements of adjustment—for businesses and workers displaced by these changes—in addition to promoting the opportunities offered by globalization. These may range from retraining programs to programs that improve the competitiveness of displaced workers and businesses.

- A strong state policy addressing trade and competitiveness is one that leverages and coordinates the resources available at other levels of government.

- Joint public-private partnerships can exploit the regional variations within a state such as California in order to promote alternative development sites that offer proximity advantages to Silicon Valley but are outside of the core area.

8.6   ISSUES AND FUTURE SCENARIOS

High-tech industries in both the US and California have greatly benefited from the process of globalization. In particular, the continued competitiveness of multinational high-tech firms and the higher incomes earned by high-tech workers has been directly linked to the availability of low-cost foreign production sites. The globalization process, however, raises longer-term issues for the state and its high-tech industries.

First, displaced manufacturing production workers are unlikely to find employment in the higher-wage, high-tech, services jobs. Developing new alternatives for these workers has frustrated policy-makers for some time. The answer, however, is not to slow the process of globalization, since the high-tech firms will then simply lose more sales to lower cost foreign producers, further reducing the job opportunities.

Second, our evidence suggests that a major benefit of globalization has been the growth in high-tech services employment that has accompanied the globalization of manufacturing production. It is not clear, however, that the economy will experience the same benefits from the accelerating globalization of high-tech services as it gained from the globalization of manufacturing. The current directions in globalization may be analyzed in terms of two alternative futures:

- One positive scenario is that the US and California economies continue to adjust to the new production paradigm, keeping the "cream" of the new development at home, while allowing the globalization of the more routine production and service activities. Under this scenario, innovation would lead to a continuing stream of new products, and hence new manufacturing and services jobs, while competition and the need for lower-cost production would force more mature manufacturing and services operations overseas.

- An alternative scenario is that the globalization of high-tech and business outsourcing services proves to be different in kind from earlier waves of globalization, with high losses of skilled white-collar jobs to lower-cost foreign sites. These moves would slow the growth of high-tech jobs within the US and California, leading to slower growth in income. As centers of skilled high-tech professionals build up in other parts of the globe, the US and California may no longer dominate the next wave of innovations, leading to the possibility of much lower long-term growth.

Whichever of these scenarios occurs, globalization will continue to shape the evolution of the high-tech industry and will continue to present businessmen, academics, and policymakers with both opportunities and challenges. The future contours of competitiveness are being determined today, and the outcome will depend significantly on how adjustments are made to the forces created by the interaction of globalization and a high-tech economy.

References

Addison, John T, Douglas A. Fox, and Christopher J. Ruhm. 1995. "Trade and Displacement in Manufacturing." *Monthly Labor Review.* 118 (4): 58-67.

Aizcorbe, Ana, Kenneth Flamm, and Anjum Khurshid. 2002. "The Role of Semiconductor Inputs in IT Hardware Price Decline: Computers vs. Communications." *Finance and Economics Discussion Series, Division of Research & Statistics and Monetary Affairs*, paper # 2002-37. Washington, D.C.: Federal Reserve Board.

Alonso, William. 1975. "Industrial Location and Regional Policy." *Regional Policy: Readings in Theory and Applications.* Editors John Friedman and William Alonso. Cambridge, MA: The MIT Press. 64-96.

————. 1975. "Location Theory." *Regional Policy: Readings in Theory and Applications.* Editors John Friedman and William Alonso. Cambridge, MA: The MIT Press. 35-63.

American Association of Engineering Societies. 2000. "Policy Statement: Ensuring a Strong, High-Tech Workforce." Web page. [Accessed 24 September 2002]. Available at <http://www.aaes.org/content.cfm?L1=2&L2=1&PID=1>.

Andersson, Thomas and Torbjorn Fredriksson. 2000."Distinction between Intermediate and Finished Products in Intra-Firm Trade". *International Journal of Industrial Organization.* 18(5): 773-792.

Anonymous. 1994. "US Export Assistance Centers: One-Stop Shops for Exporters." *Business America.* 115 (1): 2-3.

Anonymous. 1996. "Small Business." *Business America.* 117 (9): 92-109.

Anonymous. 2002. "Foreign Trade Zones: Lands of Opportunity." *Transportation & Distribution* 43 (6): 54-55.

Arndt, Sven W., and Henry Kierzkowski eds. 2001. *Fragmentation: New Production Patterns in the World Economy.* Oxford: Oxford University Press.

Bailey, Jeff. 2002. "Cuts Set in Consulting Funds for Small Manufacturers." *Wall Street Journal.* (Reprint, March 26).

Baldwin, John R. and Glen G. Cain. 1994. "Trade and U.S. Relative Wages: Preliminary Results." Mimeo. Madison: University of Wisconsin.

Baldwin, John R. and Richard E. Caves. 1998. "International Competition and Industrial Performance: Allocative Efficiency, Productive Efficiency and Turbulence." in *The Economics and Politics of International Trade: Freedom and Trade, Volume II.* ed. Gary Cook. New York: Routledge.

Baldwin, Richard and Gianmarco I. P. Ottaviano. 1998. "Multiproduct Multinationals and Reciprocal FDI Dumping." NBER Working Paper No. 6483.

Bardhan, Ashok Deo, and David K. Howe. 1998a. "Globalization and Labor: The Effect of Imported Inputs on Blue-Collar Workers." Working Paper 98-261. Berkeley: Fisher Center for Real Estate and Urban Economics.

Bardhan, Ashok Deo and David Howe. 2001. "Globalization and Restructuring During Downturns: A Case Study of California." *Growth and Change.* 32 (2): 217-235.

————. 1998b. "Transnational Social Networks, Transportation Costs, and the Geographic Distribution of California's Exports." Working Paper 98-262. Berkeley: Fisher Center for Real Estate and Urban Economics, University of California.

Bardhan, Ashok, Cynthia A. Kroll, and Dwight M. Jaffee. 1995. "The Growing Role of Foreign Trade in California's Economy." Working Paper 95-239. Berkeley: Fisher Center for Real Estate and Urban Economics, University of California.

Bartel, A. P. 1989. "Where Do the New U.S. Immigrants Live?" *Journal of Labor Economics.* 7 (4): 371-91.

Bartelsman, Eric J., Randy A. Becker, and Wayne B. Gray. "NBER-CES Manufacturing Industry Database". Available at www.nber.org/nberces/nbprod96.htm.

Belanger, A., and A. Rogers. 1992. "The Internal Migration and Spatial Redistribution of the Foreign-Born in the United States: 1965-1970 and 1975-1980." *International Migration Review.* 4: 1342-69.

Bergstrand, J. H. 1990. "The Heckscher-Ohlin-Samuelson Model, the Linder Hypothesis and the Determinants of Bilateral Intra-Industry Trade." *Economic Journal,* 100 (403): 1216-1229.

Blair, John P. 1999. "Local Economic Development and National Growth." *Economic Development Review.* 16 (3): 93-97.

———. 1991. *Urban and Regional Economics.* Boston: Richard D. Irwin, Inc.

Blanchard, O. J., L.F. Katz, R.E. Hall and B. Eichengreen. 1992. "Regional Evolutions; Comments and Discussion." *Brookings Papers on Economic Activity.* 1: 1-75.

Blanchard, Olivier Jean, Lopez-de-Silanes, Florencio and Shleifer, Andrei. 1994. "What Do Firms Do With Cash Windfalls?" *Journal of Financial Economics.* 35 (3): 337-360.

Bolton, Roger. 1992. "'Place Prosperity Vs People Prosperity' Revisited: An Old Issue with a New Angle." *Urban Studies* 29 (2): 185-203.

Borga, Maria and Michael Mann. 2002. "US International Services." *Survey of Current Business* (October): 67-124.

Borjas, George J. 1994. "The Economics of Immigration." *Journal of Economic Literature,* 32 (4): 1667-1717.

Borjas, George J. 1995. "Ethnicity, Neighborhoods, and Human-Capital Externalities." *American Economic Review.* 85 (3): 365-390.

Bovaird, Tony. 1994. "Managing Urban Economic Development: Learning to Change or the Marketing of Failure?" *Urban Studies.* 31 (4-5): 573-603.

Bowen, Harry P., Edward E. Leamer, and Leo Sveikauskas. 1987. "Multicountry Multifactor Tests of the Factor Abundance Theory." *American Economic Review* 77 (5): 791-809.

Bowman, Louise. 2002. "Irish Revival." *Strategic Direct Investor.* (March/April, 30-32).

Bradshaw, Ted K., and Edward J. Blakely. 1999. "What Are 'Third Wave' State Economic Development Efforts: From Incentives to Industrial Policy." *Economic Development Quarterly.* 13 (3): 229-44.

Brainard, S. Lael. 1997. "An Empirical Assessment of the Proximity-Concentration Tradeoff between Multinational Sales and Trade." *American Economic Review.* 87 (4): 520-544.

Braun, Alexander. 1998. "Is It the Lack or the Price of Brains?" *Semiconductor International.* 21 (13): 15.

Caballero, Ricardo J., and Mohamad L. Hammour 1994. "The Cleansing Effect of Recessions." *American Economic Review.* 84 (5): 1350-1368.

California Bureau of State Audits. 2001. *Technology, Trade and Commerce Agency: Its Strategic Planning Is Fragmented and Incomplete, and Its International Division Needs to Bet-*

ter Coordinate With Other Entities, but Its Economic Development Division Customers Generally Are Satisfied. Sacramento: California State Auditor.

California Community Colleges Economic Development Program. n.d. "Centers for Applied Competitive Technologies." Web page. [Accessed 2002]. Available at <http://www.cact. org/>.

California Council for International Trade. 2002. "California House Votes on International Trade." Web page. [Accessed 2002]. Available at www.ccit.net.

California Council on Science and Technology. 2002. "California Council on Science and Technology Home Page." Web page. [Accessed 2002]. Available at <http://www.ccst. ucr.edu/>.

California Department of Finance. 1997. "California Department of Finance Home Page." Web page. [Accessed 1997]. Available at <http://www.dof.ca.gov/>>.

————. 2002a. *California Governor's Budget 2002-2003.* Sacramento: California Department of Finance.

————. 2002b. *California Governor's Budget 2002-2003, May Revision.* Sacramento: California Department of Finance.

California Economic Strategy Panel. 1996. *Collaborating to Compete in the New Economy: An Economic Strategy for California.* Sacramento: California Trade and Commerce Agency.

————. 2000a. *Collaborating to Compete in the New Economy: Principles From the La Jolla Retreat.* Sacramento: California Economic Strategy Panel.

————. 2000b. *Collaborating to Succeed in the New Economy: Findings of the Regional Economic Development Survey* . Sacramento: California Economic Strategy Panel.

————. 1998. *Special Statewide Forum on the California Aerospace Industry Cluster, White Paper on Issues and Opportunities.* Sacramento: California Economic Strategy Panel.

California Employment Training Panel. 2002. "Employment Training Panel: Our Program." Web page. [Accessed 2002]. Available at <http://www.etp.cahwnet.gov/m_program.cfm>.

California Technology, Trade and Commerce Agency. 2002a. "Business & Community Resources: Small Businesses." Web page. [Accessed 2002]. Available at <http:// commerce.ca.gov/state/ttca/ttca_htmldisplay.jsp>.

————. 2002b. "California Technology, Trade and Commerce Agency Home Page." Web page. [Accessed 2002]. Available at <http://commerce.ca.gov/state/ttca/ttca_homepage. jsp>.

————. 2002c. "International Business." Web page. [Accessed 2002]. Available at <http:// www.commerce.ca.gov/state/ttca/ttca_navigation.jsp?path=International+Business>.

————. 2001. *International Trade and Investment, Year in Review, Fiscal Year 1999-2000.* Sacramento: California Technology, Trade and Commerce Agency.

California Trade and Commerce Agency. 1996. *International Trade and Investment, Fiscal Year 1995-96 Year in Review.* Sacramento: California Trade and Commerce Agency.

Campa, Jose and Linda Goldberg. 1997. "The Evolving External Orientation of Manufacturing Industries: Evidence from Four Countries." 1997. Federal Reserve Bank of New York. *Economic Policy Review.* 3 (2): 53-81.

Carr, David L, James R. Markusen, and Keith Maskus. 2001. "Estimating the Knowledge-Capital Model of the Multinational Enterprise." *American Economic Review.* 91 (3): 693-708.

Castles, S. and M.J. Miller. 1993. *The Age of Migration: International Population Movements in the Modern World.* London: Macmillan.

Caves, Richard E., Jeffrey A. Frankel, and Ronald W. Jones. 2001. *World Trade and Payments: An Introduction.* Boston: Addison-Wesley.

Central Intelligence Agency. 2001. *World Fact Book 2001.* Washington, D.C.: Central Intelligence Agency.

Chernotsky, Harry I. 1987. "The American Connection: Motives for Japanese Foreign Direct Investment." *Columbia Journal of World Business.* 22 (4): 47-54.

Chin, Stephen A. 1995. "Sewing Up a Model Garment Industry Agreement." *San Francisco Examiner.* (25 June, sec. B): 3.

City College of San Francisco. 2002. "Center for Applied Competitive Technologies." Web page. [Accessed 2002]. Available at http://www.cact.org/sfcact/.

Clausing, Kimberly. 2000. "Does Multinational Activity Displace Trade?" *Economic Inquiry* 38: 190-205.

Cole, Jim. 2002. "East Bay Chip Firms Set Up Shop in China." *San Francisco Business Times.* Volume 4, no. 33. (April).

Cole, Jim. 2002. "Far-Reaching Boom: Offshore Outsourcing." East Bay Business Times 5, no. 9: 9.

Conway, Carol, and William E. Nothdurft. 1996. *The International State: Crafting a State-wide Trade Development System.* Washington, D.C.: The Aspen Institute, Rural Economic Policy Program.

Cortright, Joseph and Heike Mayer. 2001. "High Tech Specialization: A Comparison of High Technology Centers." *Center on Urban and Metropolitan Policy, The Brookings Institution Survey Series* (January).

Coughlin, C., and P. Cartwright. 1987. "An Examination of State Foreign Export Promotion and Manufacturing Exports." *Journal of Regional Science.* 27 (3): 439-49.

Cox, Kevin R., Editor. 1997. *Spaces of Globalization: Reasserting the Power of the Local.* New York: The Guilford Press.

Christaller, Walter. 1933. *Die zentralen Orte in Suddeutxchland.* Translated as *Central Places in Southern Germany* by Baskin, C.W. 1966. Englewood Cliffs: Pretice-Hall.

Daly, Mary. 2002. "Riding the IT Wave: Surging Productivity Growth in the West." *Federal Reserve Bank of San Francisco Economic Letter*, no. 2002-34.

Da Ponte, John J. Jr. 1997. "The Foreign-Trade Zones Act: Keeping Up With the Changing Times." *Business America.* 118 (12): 22-25.

Davis, Steven J., John C. Haltiwanger and Scott Schuh. 1996. *Job Creation and Destruction.* Cambridge: MIT Press.

DeAngelo, Linda Elizabeth. 1988. "Managerial Competition, Information Costs, and Corporate Governance: The Use of Accounting Performance Measures in Proxy Contests." *Journal of Accounting and Economics.* 10 (1): 3-36.

DeAngelo, Harry and Linda DeAngelo. 1989. "Proxy Contests and the Governance of Publicly Held Corporations." *Journal of Financial Economics.* 23 (1): 29-59.

Deardorff, Alan V. 1984. "Testing Trade Theories and Predicting Trade Flows." *Handbook of International Economics.* Editors R. W. Jones, and P. B. Kenen, 467-517. Vol. 1. New York: North Holland/Elsevier Science Publishers.

Decker, Paul T, and Walter Corson. 1995. "International Trade and Worker Displacement: Evaluation of the Trade Adjustment Assistance Program." *Industrial and Labor Relations Review.* 48 (4): 758-74.

Deutsch, Klaus Gunter, and Bernhard Speyer, Editors. 2001. *The World Trade Organization Millennium Round.* New York: Routledge.

DeVol, Ross. 1999. "America's High-Tech Economy." Milken Institute (13 July).

Dornbusch, Rudiger, and Stanley Fischer. 1987. *Macro-Economics.* 4th Ed. New York: McGraw-Hill.

Dossani, Rafiq. 2003. "The Growth of India's IT Industry." Working Paper. Stanford: Asia Pacific Research Center.

Dunlevy, J. A. and W.K. Hutchinson. 1999. "The Impact of Immigration on American Import Trade in the Late Nineteenth and Early Twentieth Centuries." *The Journal of Economic History.* 59: 1043-62.

Egger, Peter, Michael Pfaffermayr and Yvonne Wolfmayr-Schnitzer. 2001. "The International Fragmentation of Austrian Manufacturing: The Effects of Outsourcing on Productivity and Wages." *North American Journal of Economics and Finance.* No. 12: 257–272.

Elliott, Heidi. 1998. "The Dating Game." *Electronic Business.* February: 81-83.

Empire State Development. 1997. *Business Assistance in the New Empire State.* New York: Empire State Development.

Empire State Development. 2002. "New York Loves Business." Web page. [Accessed 2002]. Available at <http://www.nylovesbiz.com/default.asp>.

Feenstra, Robert C. 1998. "Integration of Trade and Disintegration of Production in the Global Economy." *Journal of Economic Perspectives.* 12 (4): 31-50.

Feenstra, Robert C. 1996. "U.S. Imports, 1972-1994: Data and Concordances." NBER Working Paper No. 5515.

Feenstra, Robert C. and Gordon H. Hanson. 1996a. "Foreign Investment, Outsourcing and Relative Wages." *Political Economy of Trade Policy: Essays in Honor of Jagdish Bhagwati.* Robert C. Feenstra, Gene M. Grossman and Douglas A. Irwin, Editors. Cambridge, MA: MIT Press, 89-127.

Feenstra, Robert C. and Gordon H. Hanson. 1996b. "Globalization, Outsourcing and Wage Inequality." *American Economic Review.* 86 (2): 240-245. An erratum updating their tables is also available from those authors.

Feenstra, Robert C and Gordon H. Hanson. 1999. "The Impact of Outsourcing and High-Technology Capital on Wages: Estimates for the U.S., 1972-1990." *Quarterly Journal of Economics.* 114(3): 907-940.

Feenstra, Robert C., James Markusen and Andrew Rose 2001. "Using the Gravity Equation to Differentiate Among Alternative Theories of Trade." *Canadian Journal of Economics.* 34 (2): 430-447.

Foreign Trade Zone Corporation. 2002. "Foreign Trade Zone Resource Center." Web page. [Accessed September 2002]. Available at <http://www.foreign-trade-zone.com/index.html.>

Franco, Ana and Sylvain Jouhette. 2001. "Labor Force Survey: Principal Results 2000." *Statistics in Focus.* Eurostat (October).

Frankel, Jeffrey A., Ernesto Stein, and Shang-jin Wei. 1995. "Trading Blocs and the Americas: The Natural, the Unnatural, and the Super-natural." *Journal of Development Economics.* 47 (1): 61-95.

Frankel, J. A., D. Romer and T. Cyrus. 1996. "Trade and Growth in East Asian Countries: Cause and Effect?" *NBER Working Paper Series.* No. 5732.

Frankel, J. A., E. Stein and S. Wei. 1996. "Regional Trading Arrangements: Natural or Supernatural?" *American Economic Review.* 86 (2): 52-56.

Friedman, John and William Alonso, Editors. 1975. *Regional Policy: Readings in Theory and Applications.* (Cambridge: The MIT Press).

Fujita, Masahisa, Paul Krugman, and Anthony J. Venables. 2000. *The Spatial Economy* The MIT Press.

Garry, Gregory C. 1999. "Offshore Programmers: The Wave of the Future?" *UNIX Review's Performance Computing.* 17 (5): 14-19.

Gilchrist, Donald A; St Louis, Larry V. 1994. "An Equilibrium Analysis of Regional Industrial Diversification." *Regional Science & Urban Economics.* 24 (1): 115-133.

Ginsberg, Steve. 2002. "Sun Burned: Broken Lease Costs $85M." *San Francisco Business Times.* 16 (35, April 8).

Glass, Amy Jocelyn and Kamal Saggi. 2002. "Licensing versus Direct Investment: Implications for Economic Growth." *Journal of International Economics.* 56 (1): 131-153.

Gould, D. 1994. "Immigrant Links to the Home Country: Empirical Implications for U.S. Bilateral Trade Flows." *Review of Economics & Statistics.* 76 (2): 302-316.

Greising, David. 1988. "It's the Best of Times—Or Is It?" *Business Week.* Jan 12.

Grossman, Gene and Elhanan Helpman 2002. "Outsourcing in a Global Economy," *NBER Working Paper No. 8728*

Hackbarth, Amy. 2002. "Finding an Off-Shore Work Pool." *Minneapolis-St Paul Business Journal.*

Head, C. Keith, John C. Ries, and Deborah L. Swenson. 1999. "Attracting Foreign Manufacturing: Investment Promotion and Agglomeration." *Regional Science and Urban Economics.* 29 (2): 197-218.

Head, K. and J. Ries. 1998. "Immigration and Trade Creation: Econometric Evidence from Canada." *Canadian Journal of Economics.* 31 (1): 47-62.

Helpman, Elhanan. 1984. "A Simple Theory of International Trade with Multinational Corporations." *Journal of Political Economy.* 92 (3): 451-471.

Helpman, Elhanan, and Paul R. Krugman. 1985. *Market Structure and Foreign Trade: Increasing Returns, Imperfect Competition, and the International Economy.* (Cambridge: MIT Press).

Hewlett Packard. n.d. "HP History and Facts." Web page. [Accessed 2002]. http://www.hp.com/hpinfo/abouthp/histnfacts/index.htm.

Hildebrand, George, and Arthur Mace Jr. 1950. "The Employment Multiplier in an Expanding Industrial Market: Los Angeles County, 1940-47." *Review of Economics and Statistics* XXXII (1950): 341-349.

Hirschman, Albert O. 1958. "Interregional and International Transmission of Economic Growth." *The Strategy of Economic Development.* (New Haven: Yale University Press).

As reprinted in *Regional Policy: Readings in Theory and Applications*. Editors John Friedman, and William Alonso. Cambridge, MA: The MIT Press. 139-157.

Ho, C. G. T. 1993. "The Internationalization of Kinship and the Feminization of Caribbean Migration: The Case of Afro-Trinidadian Immigrants in Los Angeles." *Human Organization*. 52: 32-40.

Holdway, Michael. 2001. "Quality-Adjusting Computer Prices in the Producer Price Index: An Overview." Bureau of Labor Statistics, US Department of Labor. (16 October); Available at <http://www.bls.gov/ppi/ppicomqa.htm>.

Holland, Stuart. 1976. *Capital Versus the Regions*. New York: St. Martin's Press.

Hoover, Edgar M. 1971. *An Introduction to Regional Economics*. New York: Alfred A. Knopf.

Hoover's Guide to Computer Companies. 1996. Second Edition. Austin: Hoover's Business Press.

Hoover's Online. Web page. [Accessed 2002 and 2003]. Available at <http://www.hoovers.com/>.

Howes, Candace, and Ann R. Markusen. 1993. "Trade, Industry, and Economic Development." In *Trading Industries, Trading Regions: International Trade, American Industry, and Regional Economic Development*. Editors Helzi Naponen, Julie Graham, and Ann R. Markusen. New York: The Guilford Press. 10-44.

Hummels, David, Dana Raporport, and Kei-Mu Yi. 1997. "Globalization and the Changing Nature of World Trade." *Federal Reserve Bank of New York, Economic Policy Review*. 3 (2): 53-81.

International Labor Organization. "LABORSTA." Web page. [Accessed 2003]. Available at <http://laborsta.ilo.org>.

Irwin, Douglas. 1996. "The United States in a New World Economy? A Century's Perspective." *American Economic Review*. 86 (2): 41-51.

Isard, Walter. 1956. *Location and Space Economy*. New York: The Technology Press and John Wiley & Sons, Inc.

Jaffee, Dwight M. 1998a. "International Trade and California's Economy: Summary of the Data." Working Paper 98-258. Berkeley: Fisher Center for Real Estate and Urban Economics, University of California.

Jaffee, Dwight M. 1998b. "International Trade and California Employment: Some Statistical Tests." Working Paper 98-259. Berkeley: Fisher Center for Real Estate and Urban Economics, University of California.

Johansson, Helena, and Lars Nilsson. 1997. "Export Processing Zones As Catalysts." *World Development*. 25 (12): 2115.

Joint Venture Silicon Valley Network. n.d. Web page. [Accessed 2002]. Available at <http://www.jointventure.org/>.

Kansas Department of Commerce and Housing. 2002. "International Business." Web page. [Accessed 2002]. Available at <http://kdoch.state.ks.us/busdev/trade_info.jsp>.

Katz, Arnold and Shelby W. Herman. 1997. "Improved Estimates of Fixed Reproducible Tangible Wealth 1929-95." *Survey of Current Business*. (May): 69-92.

Kemp, Murray C. and Koji Shimomura. 1999 "The Internationalization of the World Economy and its Implications for National Welfare." *Review of International Economics*. 7 (1, February): 1-7.

Khirallah, Diane Rezendes. 2002. "Where Does H-1B Fit?" *InformationWeek*. 874 (February 4). 34-42.

Kim, Sukkoo. 1995. "Expansion of Markets and the Geographic Distribution of Economic Activities: The Trends in U.S. Regional Manufacturing Structure, 1860-1987." *Quarterly Journal of Economics*. 110 (4): 881-908.

Knarvik, Karen Helene Midelfart, Tvedt, Jostein. 2000. "International Trade, Technological Development, and Agglomeration." *Review of International Economics* 8 (1): 149-163.

Knowles Mathur, Lynette, and Ike Mathur. 1997. "The Effectiveness of the Foreign-Trade Zone As an Export Promotion Program: Policy Issues and Alternatives." *Journal of Macromarketing*. 17 (2): 20-31.

Koehler, Gus A. 1994. *New Challenges to California State Government's Economic Development Engine*. Sacramento: California State Library, California Research Bureau.

Koehler, Gus. 1999. *California Trade Policy*. Sacramento: California State Library, California Research Bureau.

Koehler, Gus, and Costolino Hogan. 1996. *State Government Economic Development Programs*. Sacramento: California State Library, California Research Bureau.

Konan, Denise Eby. 2000. "The Vertical Multinational Enterprise and International Trade." *Review of International Economics*. 8 (1): 113-125.

Krause, Peggy. 1993. "Can Your Business Profit in an FTZ?" *Global Trade & Transportation* 113 (6): 11.

Kroll, Cynthia A. 1998b. "Foreign Trade and California's Economic Growth: A Summary of Findings and Directions for Policy." Working Paper 98-263. Berkeley: Fisher Center for Real Estate and Urban Economics, University of California.

Kroll, Cynthia, and Ashok Bardhan. 1996. "Examining the Role of Travel and Tourism in California's Economy." *Quarterly Report*. Fisher Center for Real Estate and Urban Economics, University of California. Spring: 1-9.

Kroll, Cynthia A., Dwight M. Jaffee, Ashok Deo Bardhan, Josh Kirschenbaum, and David K. Howe. 1998. *Foreign Trade and California's Economic Growth*. Berkeley: California Policy Seminar, University of California.

Kroll, Cynthia A., and Josh Kirschenbaum. 1998. "The Integration of Trade into California Industry: Case Studies of the Computer Cluster and the Food Processing Industry." Working Paper 98-260. Berkeley: Fisher Center for Real Estate and Urban Economics, University of California.

Krueger, Anne. 2000. "NAFTA's Effects: A Preliminary Assessment." *The World Economy*; 23 (6): 761-775.

Krugman, Paul. 1991. *Geography and Trade*. Cambridge: MIT Press.

———. 1994. *Peddling Prosperity*. New York: W.W. Norton & Company.

Krugman, Paul and Anthony J. Venables. 1996. "Integration, Specialization, and Adjustment." *European Economic Review* 40 (3-5): 959-967.

Kuemmerle, Walter. 1999. "The Drivers of Foreign Direct Investment into Research and Development: An Empirical Investigation." *Journal of International Business Studies*. 30 (1): 1-24.

Lamertz, Kai and Joel Baum. 1998. "The Legitimacy of Organizational Downsizing in Canada: An Analysis of Explanatory Media Accounts." *Canadian Journal of Administrative Sciences*. 15: 93-107.

Larson, Eric M., Charles A. Moore, Patrick C. Seeley, and Angela S. Bourciquot. 1992. *Immigration and the Labor Market: Nonimmigrant Alien Workers in the United States.* GAO/PEMD-92-17. Washington, D.C.: US General Accounting Office.

Lieberman, Marvin B. 1991. "Determinants of Vertical Integration: An Empirical Test." *Journal of Industrial Economics.* 39 (5): 451-466.

Lingblom, Marie. 2002. "Outsourcing Firms in India: Time to Move on?" *CRN: The Newsweekly for Builders of Technology.* 1001: 33.

Lösch, August. 1938. "The Nature of Economic Regions." *Southern Economic Journal.* 5 (1): 71-88. Reprinted in *Regional Policy: Readings in Theory and Applications.* Editors John Friedman, and William Alonso. Cambridge, MA: The MIT Press, 1975. 97-105.

Madan, Vibhas. 2000. "Transfer Prices and the Structure of Intra-firm Trade." *Canadian Journal of Economics.* 33 (1): 53-68.

Mailman, Stanley. 1995. "California Proposition 187 and Its Lessons." *New York Law Journal.* 3.

Malecki, E.J. 1979. "Locational Trends in R&D by Large U.S. Corporations, 1965-1977." *Economic Geography.* 55 (4): 309-322.

_____. 1985. "Industrial Location and Corporate Organization in High Technology Industries." *Economic Geography.* 61 (4): 345-369.

Markusen, A. 1985. *Profit Cycle, Oligopoly, and Regional Development.* Cambridge: MIT Press.

Markusen, Ann. 1994. "Interaction Between Regional and Industrial Policies: Evidence From Four Countries." *The World Bank Research Observer.* 279-298.

Markusen, Ann R. 1993. "Trade As a Regional Development Issue: Policies for Job and Community Preservation." *Trading Industries, Trading Regions: International Trade, American Industry, and Regional Economic Development.* Editors Helzi Naponen, Julie Graham, and Ann R. Markusen. New York: The Guilford Press. 285-302.

Markusen, Ann, Peter Hall and Amy Glasmeier. 1986. *High Tech America.* Boston: Allen & Unwin.

Markusen, Ann, Helzi Naponen, and K. Driessen. 1991. "International Trade, Productivity and Regional Growth." *International Regional Science Review* 14 (1): 15-39.

Markusen, James R. 1995. "The Boundaries of Multinational Enterprises and the Theory of International Trade," *Journal of Economic Perspectives*, 9(2), 169-189.

Markusen, James R. and Keith Maskus. 2001. "Multinational Firms: Reconciling Theory and Evidence." *Topics in Empirical International Economics: A Festschrift in Honor of Robert E. Lipsey.* Magnus Blomström and Linda S. Goldberg, Editors. Chicago: The University of Chicago Press.

Markusen, James, and James Melvin. 1988. *The Theory of International Trade.* New York: Harper & Row.

Markusen, James and Anthony J. Venables. 1995. "Multinational Firms and the New Trade Theory." *NBER Working Paper No. 5036*

Maryland Department of Business and Economic Development. 2002. "International Business." Web page. [Accessed 2002]. Available at http://www.choosemaryland.org/ international/index.asp.

Massachusetts Institute for Social and Economic Research. 2002. "Foreign Trade Database." Web page. [Accessed 2002]. Available at http://misertrade.org/.

Massey, D. S. and F.G. España. 1987. "The Social Process of International Migration." *Science*. 237: 733-738.

Mataloni, Jr., Raymond J. 1999. "U.S. Multinational Companies: Operations in 1997." *Survey of Current Business*. July: 8-35.

Matthei, L. M. 1996. "Gender and International Labor Migration: A Networks Approach." *Social Justice*. 23 (3): 38-53.

Mayer, John. 2002. "Redefining Ireland--After Decades of Success As a Low-Cost Haven for Electronics Firms, the Emerald Isle May Have to Forge an New Identity." *EBN*. 1327 (August 26): 27.

Microsoft Encarta World Atlas. 2000.

Miller, J.P. 1989. "The Product Cycle and High Technology Industry in Non-Metropolitan Areas, 1976-1980." *The Review of Regional Studies*. 19 (1): 1-12.

Mittelman, James H. 2000. *The Globalization Syndrome: Transformation and Resistance*. Princeton: Princeton University Press.

Nair-Reichert, Usha and Diana Weinhold. 2001. "Causality Test for Cross-Country Panels: A New Look at FDI and Economic Growth in Developing Countries." *Oxford Bulletin of Economics & Statistics*. 63 (2): 153-172.

Naponen, Helzi, Julie Graham, and Ann R. Markusen. 1993. *Trading Industries, Trading Regions: International Trade, American Industry, and Regional Economic Development*. New York: The Guilford Press, 1993.

National Association of Software and Services Companies. "NASSCOM." Web page. [Accessed 2003]. Available at <http://www.nasscom.org>.

National Association of State Development Agencies. 2002. "Welcome to NASDA!" Web page. [Accessed 2002]. Available at <http://www.nasda.com/>.

National Governors Association. 2002. *A Governor's Guide to Trade and Global Competitiveness*. Washington, D.C.: National Governors Association.

National Institute of Standards and Technology. 2002. "Budget, Planning and Economic Analysis: Technology Administration FY 2003 Budget Overview." Web page. [Accessed 2002]. Available at <http://www.nist.gov/public_affairs/budget/2003overview.htm>.

Nobre, Ana. 1999. "Labor Costs 1996: Major Disparities between the European Union Countries." *Statistics in Focus*.

Nothdurft, William E. 1992. *Going Global: How Europe Helps Small Firms Export*. Washington, D.C.: The Brookings Institution.

North, Douglas C. 1955. "Location Theory and Regional Economic Growth." *Journal of Political Economy*. 63 (3): 243-258. Reprinted in *Regional Policy: Readings in Theory and Applications*. Editors John Friedman and William Alonso. Cambridge, MA: The MIT Press, 1975. 332-347.

Norton, Leslie P. 2002. "Asian Trader: Investors Listen Once More to the Song of India." *Barron's*. 82 (49): 12. (December 9).

Noussair, Charles N., Charles R. Plott, and Raymond G. Riezman. 1995. "An Experimental Investigation of the Patterns of International Trade." *American Economic Review*. 85 (3): 462-491.

Noyelle, Thierry J., and Thomas M. Stanback Jr.. 1984. *The Economic Transformation of American Cities*. Totowa: Rowman & Allanheld.

Oklahoma Department of Commerce. 2002. "International Trade and Investment." Web page. [Accessed 2002]. Available at <http://domino1.odoc.state.ok.us/Menu4.nsf/GlobalFrame>.

OECD. 2002. "Intra-Industry and Intra-Firm Trade and the Internationalization of Production." *Economic Outlook.* 71: Chapter VI.

Oliner, Stephen and Daniel Sichel. 2002. "Information Technology and Productivity: Where Are We Now and Where Are We Going?" Working Paper 2002-29. Washington, D.C.: Board of Governors of the Federal Reserve System, Finance and Economics Discussion Series.

Orbuch, Paul M., and Thomas O. Singer. 1995. "International Trade, the Environment, and the States: An Evolving State-Federal Relationship." *Journal of Environment and Development.* 4 (2): 121-144.

Osterman, Paul. 1993. "Pressures and Prospects for Employment Security in the United States." *Employment Security and Labor Market Behavior.* Christoph F. Buechtemann, Editor. Ithaca: ILA Press. 228-243.

Palmer, Ian, Boris Kabanoff and Richard Dunford. 1997. "Managerial Accounts of Downsizing." *Journal of Organizational Behavior.* 18 (S1): 623-639.

Parker, Eric. 1997. *Regional Industrial Revitalization: Implications for Workforce Development Policy.* Working Paper 114. New Brunswick: Project on Regional and Industrial Economics.

Parks, Elizabeth. 1998. "Is America Really Low on High-Tech Workers?" *Machine Design.* 70 (16): 70-74.

Porter, Michael. 2000. "Location, Competition, and Economic Development: Local Clusters in a Global Economy." *Economic Development Quarterly.* 14 (1): 15-34.

Posner, Alan R. 1984. *State Government Export Promotion: An Exporter's Guide.* Westport: Quorum Books.

Prasch, Robert E. 1995. "Reassessing Comparative Advantage: The Impact of Capital Flows on the Argument for Laissez-Faire." *Journal of Economic Issues.* 29 (2): 427-433.

Rauch, James. 1999. "Networks Versus Markets in International Trade." *Journal Of International Economics.* 48 (1): 7-35.

Rauch, James. 2001. "Business and Social Networks in International Trade." *Journal of Economic Literature.* 39 (4): 1177-1203.

Renner, G.T. 1947. "Geography of Industrial Localization." *Economic Geography.* 23: 167-189.

Ricci, Luca Antonio. 1999. "Economic Geography and Comparative Advantage: Agglomeration versus Specialization." *European Economic Review.* 43 (2): 357-377.

Richardson, Harry W. 1977. *Regional Growth Theory.* London: MacMillan Press Ltd.

Richardson, J. David, Geza Feketekuty, Chi Zhang, and A. E. Rodriguez. 1998. "U.S. Performance and Trade Strategy in a Shifting Global Economy." In *Trade Strategies for a New Era.* Geza Feketekuty and Bruce Stokes, Editors. New York: Council on Foreign Relations. 39-64.

Riche, Richard W., Daniel E. Hecker and John U. Burgan. 1983. "High Technology Today and Tomorrow: A Small Slice of The Employment Pie." *Monthly Labor Review.* 106 (11, November): 50-58.

Riches, Erin , Delaine McCullough, and Jean Ross. 2002. *Maximizing Returns: A Proposal for Improving the Accountability of California's Investments in Economic Development.* Sacramento: California Budget Project.

Rigby, David L and Jurgen Essletzbichler. 1997. "Evolution, Process Variety, and Regional Trajectories of Technological Change in U.S. Manufacturing." *Economic Geography.* 73 (3): 269-284.

Riker, David and S. Lael Brainard. March 1997. "US Multinationals and Competition from Low Wage Countries." NBER Working Paper No. 5959.

Rolfe, Robert J., David A. Ricks, Martha M. Pointer, and Mark McCarthy. 1993. "Determinants of FDI Incentive Preferences of MNEs." *Journal of International Business Studies.* 24 (2): 335-355.

Rosen, Howard. 1998. "Workforce Training: Investing in Human Capital or Antidote to International Trade?" In *Trade Strategies for a New Era.* Geza Feketekuty, and Bruce Stokes, Editors. New York: Council on Foreign Relations. 83-98.

Rotemberg, Julio J. and Garth Saloner. "Competition and Human Capital Accumulation: A Theory of Interregional Specialization and Trade." *Regional Science & Urban Economics.* 30 (4): 373-404.

Roy, Santanu and Jean-Marie Viaene. 1998. "On Strategic Vertical Foreign Investment." *Journal of International Economics.* 46 (2): 253-279.

Rugman, Alan M., and Gavin Boyd, Editors. 2001. *The World Trade Organization in the New Global Economy.* Northampton, MA: Edward Elgar.

Sampson, Gary P., Editor. 2001. *The Role of the World Trade Organization in Global Governance.* New York: United Nations University Press.

San Francisco Business Times. 2000. "Oracle Sees Opportunities in France." *San Francisco Business Times* Web page. (January).

———. 2000. "Oracle to Support Incubators in Thailand." *San Francisco Business Times* Web page. (August).

———. 2000. "Oracle Strikes Deal with Vietnam Trade Officials." *San Francisco Business Times* Web page. (November).

Sassen, S. 1988. *The Mobility of Labor and Capital: A Study in International Investment and Labor Flow.* London: Cambridge University Press.

Sassen, S. 1998. *Globalization and its Discontents: Essays on the New Mobility of People and Money.* New York: New Press.

Saxenian, AnnaLee. 1994. *Regional Advantage: Culture and Competition in Silicon Valley and Route 128.* Cambridge: Harvard University Press.

Saxenian, AnnaLee. 1999. *Silicon Valley's New Immigrant Entrepreneurs.* San Francisco: Public Policy Institute of California.

Saxenian, AnnaLee, and Jumbi Edulbehram. 1998. "Immigrant Entrepreneurs in Silicon Valley." *Berkeley Planning Journal.* 12: 32-49.

Saxenian, AnnaLee, Yasuyuki Motoyama, and Xiaohong Quan. 2002. *Local and Global Networks of Immigrant Professionals in Silicon Valley.* San Francisco: Public Policy Institute of California.

Scouton, William O. 1989. "States See Exports as Tool for Local Business Expansion." *Business America.* 110 (4, February 27): 7-11.

Sherman, Stratford. 1993. "Andy Grove: How Intel Makes Spending Pay Off." *Fortune.* February 22: 56-58.

Shove, Christopher. 1996. "A Simplified, Globally Competitive Economic Development Policy Framework." *Economic Development Review.* 14 (2): 10-13, (Spring).

Sihag, B. S. and C.C. McDonough. 1989. "Shift-share Analysis: The International Dimension." *Growth & Change.* 20 (3): 80-88.

Simon, J. L. 1992. "Immigrants and Alien Workers." *Journal of Labor Research.* 13 (1): 73-78.

Smith, David M. 1971. *Industrial Location: An Economic Geographical Analysis.* New York: John Wiley & Sons, Inc.

Sommers, Paul and Daniel Carlson. 2000. "Ten Steps to a High Tech Future: The New Economy in Metropolitan Seattle." Discussion Paper Prepared for the Brookings Institution Center on Urban and Metropolitan Affairs, Washington, D.C. (20 December). This paper is currently available at: http://www.brook.edu/dybdocroot/urban/sommers/ sommersexsum.htm.

South Carolina Department of Commerce. 2002. "Welcome to the South Carolina Department of Commerce." Web page. [Accessed 2002]. Available at http://www.callsouthcarolina.com/.

Sprouse, Martin, Editor. 1992. *Sabotage in the American Workplace: Anecdotes of Dissatisfaction, Mischief, and Revenge,* San Francisco: Pressure Drop Press.

Stiroh, Kevin. 2002. "Information Technology and the U.S. Productivity Revival: What do the Industry Data Say?" American Economic Review. 92 (5): 1559-1576. .

Subramanian, Rangan and Robert Z. Lawrence. 1999. "Search and Deliberation in International Exchange: Learning from Multinational Trade about Lags, Distance Effects and Home Bias." NBER Working Paper No. 7012.

Swenson, Deborah L. 2000. "Firm Outsourcing Decisions: Evidence from U.S. Foreign Trade Zones." *Economic Inquiry.* 38 (2): 175-189.

Taylor, Alex III. 1994. "The Auto Industry Meets the New Economy." *Fortune.* September 5: 52-60.

Taylor, Virginia Anne. 2002. "Analytic Framework for Global Transfer-Pricing." *Journal of American Academy of Business.* March: 308-313.

Tennessee Department of Economic and Community Development. 2002. "Office of International Affairs." Web page. [Accessed 2002]. Available at http://www.state.tn.us/ecd/idg.htm.

Thomas, Denny, and Anshuman Daga. 2002. "Outsourcing Boom Pumps Up Indian Tech Shares." *Rediff.Com Money Matters.*

Thomas, Ward F., and Paul Ong. 2002. "Barriers to Rehiring of Displaced Workers: A Study of Aerospace Engineers in California." *Economic Development Quarterly.* 1 (2): 167-178.

Thompson, Wilbur. 1958. "Internal and External Factors in the Development of Urban Economies ." In *Issues in Urban Economics.* Perloff, Harvey S. and Lowdon Wingo, Jr., Ed. Baltimore: Johns Hopkins Press. Reprinted in *Regional Policy: Readings in Theory and Applications.* Editors John Friedman, and William Alonso. Cambridge, MA: The MIT Press, 1975. 201-220.

Tiebout, Charles. 1962. *The Community Economic Base Study.* New York: The Committee for Economic Development.

Tiebout, Charles M. 1956. "Exports and Regional Economic Growth." *Journal of Political Economy.* 64 (2): 160-169. Reprinted in *Regional Policy: Readings in Theory and Applications.* Editors John Friedman, and William Alonso, 348-356. Cambridge, MA: The MIT Press, 1975.

Turnvey, William H. and Daniel C. Feldman. 1998. "Psychological Contract Violations During Corporate Restructuring." *Human Resource Management.* 37 (1): 71-83.

Tyson, Laura D'Andrea. 1987. "Creating Advantage: Strategic policy for national competitiveness." Working Paper 23. Berkeley Roundtable on the International Economy, University of California, Berkeley.

UNESCO Institute for Statistics. 2001. "The State of Science and Technology in the World." Montreal: UNESCO Institute for Statistics.

US Bureau of Economic Analysis. 2002. "Regional Accounts Data." Web page. [Accessed 2002]. Available at http://www.bea.doc.gov/bea/regional/data.htm.

————. 1997. *Survey of Current Business.* US Department of Commerce. (February).

US Bureau of the Census. 2002a. *Annual Survey of Manufactures 2000.* Available at http://www.census.gov/mcd/asmhome.html. Earlier years also available at this web site.

————. 2000. *Economic Census.* Available at <http://www.census.gov/epcd/ www/econ97. html>.

————. 1997. *Economic Census, Census of Manufactures, 1992.* Available at <http://www.census.gov/prod/1/manmin/92mmi/>.

————. 2002b. "Statistical Abstract of the United States 2001." Web page. Available at <http://www.census.gov/prod/2002pubs/01statab/stat-ab01.html>.

US Central Intelligence Agency. 2001. *World Factbook 2001.* Most recent version available at <http://www.cia.gov/cia/publications/factbook/>.

US Department of Commerce. 2002. BISNIS web site. Available at <http://www.bisnis.doc. gov/>.

US Department of Commerce, International Trade Administration. 2002. "State Exports to Countries and Regions." Web page. [Accessed 2002]. Available at <http://www.ita.doc. gov/td/industry/otea/state>.

US General Accounting Office. 1992. "Immigration and the Labor Market: Nonimmigrant Alien Workers in the United States." Report to the Chairman, Subcommittee on Immigration and Refugee Affairs, Committee on the Judiciary, US Senate. Washington D.C. (April).

US Immigration and Naturalization Service. 2000. "Characteristics of Specialty Occupation Workers (H-1B) May 1998 to July 1999." (February).

————. 2002a. "Report on Characteristics of Specialty Occupation Workers (H-1B): Fiscal Year 2001." (July).

US Immigration and Naturalization Service. 2002b. "H-1B: Specialty Occupation Workers Statistical Reports." Web page. Available at <http://www.ins.usdoj.gov/graphics/services/ employerinfo/h1b.htm>.

US Immigration and Naturalization Service. 2003. "Temporary Admissions," *Statistical Year Book of the Immigration and Naturalization Service 2000.* Available at <http://www. immigration.gov/graphics/aboutus/statistics/00yrbk_TEMP/Temp2000.pdf>.

US International Trade Commission. 2002. "Interactive Tarriff and Trade Data Web." [Accessed 2002]. <http://dataweb.usitc.gov>.

VARITA--Virtual Alliance of Russian IT Associations. 2003. "India Buys Russian Programmers." Web page. (February). Available at <http://www.russia-software.com/news.asp?filterID_News=158>.

Vijayan, Jaikumar and Thomas Hoffman. 2002. "IT Pros Prepare As War in India Looms." *Computerworld*. 36 (24): 1,16.

Vinson, Robert and Paul Harrington. 1979. "Defining High-Tech Industry in Massachussetts." Boston: Massachusetts Department of Manpower Development.

Walz, Uwe. 1996. "Transport Costs, Intermediate Goods, and Localized Growth." *Regional Science & Urban Economics*. 26 (6): 671-695.

Warner, Jerold B., Ross L. Watts and Karen H. Wruck. 1988. "Stock Prices and Top Management Changes." *Journal of Financial Economics*. 20: 461-492.

Weber, Alfred. 1929. *Alfred Weber's Theory of the Location of Industries*. (Chicago: University of Chicago Press).

Weststart - CALSTART. 2002. "Transforming Transportation." Web page. [Accessed 2002]. Available at <http://www.calstart.org/calindex3.html>.

Wilamoski, Peter and Sarah Tinkler. 1999. "The Trade Balance Effects of U.S. Foreign Direct Investment in Mexico." *Atlantic Economic Journal*. 27: 24-37.

Wilkerson, Chad. "How High Tech is the Tenth District." *Economic Review of the Federal Reserve Bank of Kansas City* 87 (2, Second Quarter): 27-54.

Wilkinson, Timothy J. 1999. "The Effect of State Appropriation on Export-Related Employment in Manufacturing." *Economic Development Quarterly*. 13 (2): 172-182.

Williamson, O. E. 1998. "Transaction Cost Economics: How it Works; Where it is Headed." *Economist-Leide*. 146 (1): 23-58.

Wilson, Daniel. 2002. "Productivity in the Twelfth District." *Federal Reserve Bank of San Francisco Economic Letter*, no. 2002-33.

Winnick, Louis. 1966. "Place Prosperity Vs. People Prosperity: Welfare Considerations in the Geographic Distribution of Economic Activity ." *Essays in Urban Land Economics*. Los Angeles: University of California, Real Estate Research Program.

World Bank. 2002. *Globalization, Growth and Poverty*. Washington, D.C.: World Bank and Oxford University Press.

World Bank. "World Development Indicators." Web page. [Accessed 2003]. Available at www. worldbank.org/data/dataquery.html.

Zedillo, Ernesto. 2003. "Will the Doha Round Implode in 2003?" *www.Forbes.com/Current Events*.

Zeile, William J. 1997. "U.S. Intrafirm Trade in Goods." *Survey of Current Business*. February: 23-38.

Zeile, William J. 1999. "Foreign Direct Investment in the United States: Preliminary Results From the 1997 Benchmark Survey." *Survey of Current Business*. August: 21-54.

Zhang, Kevin Honglin. 2001. "Does Foreign Direct Investment Promote Economic Growth? Evidence from East Asia and Latin America." *Contemporary Economic Policy*. 19 (2): 175-185.

Subject Index

Author Index